Protecting Suburban America

Protecting Suburban America

Gentrification, Advocacy, and the Historic Imaginary

Denise Lawrence-Zúñiga

Routledge
Taylor & Francis Group

LONDON AND NEW YORK

First published 2016 by Bloomsbury Academic

2 Park Square, Milton Park, Abingdon, Oxfordshire OX14 4RN
52 Vanderbilt Avenue, New York, NY 10017

Routledge is an imprint of the Taylor & Francis Group, an informa business

First issued in paperback 2020

British Library Cataloguing-in-Publication Data
A catalogue record for this book is available from the British Library.

Library of Congress Cataloging-in-Publication Data
Names: Lawrence-Zúñiga, Denise, author.
Title: Protecting suburban America : gentrification, advocacy and the historic imaginary / Denise Lawrence-Zúñiga.
Description: New York : Bloomsbury Academic, 2016.
Identifiers: LCCN 2015043991| ISBN 9781474240819 (hardback) | ISBN 9781474240826 (epub)
Subjects: LCSH: Suburban homes--Conservation and restoration--Social aspects--United States. | Historic preservation--Social aspects--United States. | Architecture--Human factors--United States. | Sociology, Urban--United States. | BISAC: ARCHITECTURE / Historic Preservation / General. | ARCHITECTURE / Urban & Land Use Planning. | SOCIAL SCIENCE / Anthropology / Cultural. | ARCHITECTURE / History / Modern (late 19th Century to 1945).
Classification: LCC NA7571 .L39 2016 | DDC 720/.47--dc23 LC record available at http://lccn.loc.gov/2015043991

Typeset by Fakenham Prepress Solutions, Fakenham, Norfolk NR21 8NN

ISBN 13: 978-1-4742-4081-9 (hbk)
ISBN 13: 978-0-367-66843-3 (pbk)

To Richard

Contents

List of Illustrations   viii

Acknowledgments   ix

1   Framing Preservation   1

2   Discovering Material Agency: Making the Preservation Homeowner   23

3   Restoration Strategies, Imagining the Past, and Reconstructing
    Historic Meaning   49

4   Historic Preservation as Cosmology: Municipal Regulations and
    City Dynamics   79

5   Local Level Preservation and Exclusion: Traditional Elites   103

6   The Gentry Move In: Education, Reform, and Advocacy   121

7   Immigrant Challenges: Communicating Preservation Values across
    the Cultural Divide in Alhambra   139

8   Toward an Anthropology of the Protected Suburb   159

Appendix A: Historic Preservation Interview Questions   171

Appendix B: Historic Preservation Questions for City Officials   175

Notes   179

References   181

Index   189

List of Illustrations

Figure 1. Traditional Craftsman house — 3

Figure 2. Traditional Spanish Revival house — 4

Figure 3. Map of southern California region indicating five cities and the Asian settlement area, the "ethnoburb" — 19

Figure 4. Restored fireplace — 29

Figure 5. Built-in china cabinet — 30

Figure 6. Old oak front door — 31

Figure 7. The celebrated multiple garages — 39

Figure 8. De-stuccoing around the doggie door — 41

Figure 9. Shingles restored on de-stuccoed house — 42

Figure 10. "Ghost" from previous construction — 45

Figure 11. Kitchen remodeled with shiplap applied to the new refrigerator to the right — 57

Figure 12. Kitchen showing original cabinets without toe-kick and remodeled new cabinets to the right with a toe-kick — 65

Figure 13. Restored wallpaper "bump" — 69

Figure 14. Photograph of house before it was remodeled — 75

Figure 15. House remodeled to "original" condition — 76

Figure 16. Residential district marker, Ontario — 95

Figure 17. Home tour in Cottage Heights — 131

Figure 18. A new mansion surrounded by single story bungalows — 141

Figure 19. Alhambra Residential Design Guidelines: Craftsman Style — 145

Figure 20. Bungalow undergoing remodeling — 152

Acknowledgments

The impetus for this qualitative study of residential historic preservation began in 2004 in the context of a course I was teaching on the subject to architecture students at my university. A nearby city was concerned about a lack of homeowner compliance with historic preservation regulations, especially in some of the lower-income neighborhoods. An increasing number of early twentieth-century wood-sided bungalows were being covered with stucco as a way to modernize and improve the properties, and make them easier to maintain. City officials, however, were worried that new owners were destroying the historic character of their homes. While the city and preservation advocates saw the houses as representations of the city's historic past, these owners seemed to see their houses as serviceable shelter. City planners wanted to know how they could educate residents to not stucco their bungalows, so as to preserve them. Around the same time, the city was reviewing its historic resource database of over 3,000 properties that had been first surveyed in 1983. Because so many houses on the list had been irreversibly remodeled in the interim, mostly using stucco to cover the wood siding, the city was working on "de-listing" those properties. In addition, planners hoped to avoid further alterations to what remained of a traditional residential landscape. The question about preservation that first stimulated this project, however, is not why people stucco old houses, but why people cared so much about preserving the original material qualities. What motivated their worry about materiality and their advocacy for convincing people to change their modernizing ways?

The "stucco" question is one that resonates with many preservation and planning professionals and design educators, but it is often less apparent to these actors why there should be a question about conserving artifacts at all. The impulse to preserve seems "natural" to professionals who work in the arts and architecture fields, so that asking questions of colleagues in design and planning often triggers deep reflection. This book is the result of ten years of research and contemplation stimulated in the context of teaching in an architecture department and conversations with colleagues, especially those with specializations in historic preservation. I owe particular debts to Lauren Bricker, Luis Hoyos, Gary McGavin, Ana Maria Whitaker, and Julianna Delgado, who shared with me their expert knowledge and practice in historic preservation throughout southern California, and generously provided contacts, information, and an often-needed sounding board for my earliest inquiries. My deepest debt goes to Daniel Lawrence for introducing me to architecture, architecture history, and professional practice, and for also helping a somewhat naive anthropologist acquire basic knowledge and achieve competent understanding.

This project was helped by advocates and residents in the five cities and throughout the region along the way. In Pasadena, where I also live, residents on my street, and on the street behind my house, have been active in historic preservation, largely to stop demolitions and mansion construction that they perceived as neighborhood threats. Historic preservation advocates in Pasadena, including Claire Bogaard and Julianna Delgado, provided valuable information on local history and introductions to possible interviewees in city hall and also the community at large. In Monrovia, I was fortunate to be assisted by Dana Hendrix and Jim Wigton of the Monrovia Old House Preservation Group. In Ontario, I worked through Cathy Walstrom in the planning department, who helped me with my initial stucco project, and Richard Delman, who has recently become an important leader in Ontario Heritage. Janet Hansen, Deputy Manager of the City of Los Angeles Office of Historic Resources, provided insightful information about the cities of Ontario and Riverside. Lauren Bricker provided an introduction to Marion Mitchell-Wilson in Riverside, who provided critical insight about the transition of the city's Cultural Heritage program, from the public library to the planning department. Kathy Maddox of the Old Riverside Foundation provided insight into Riverside's historic preservation traditions, and Justin Scott-Coe of Riverside's Green Team facilitated many introductions to homeowners. Finally, Joyce and Oscar Amaro, and Chris Olson and Lisa Selje of the Alhambra Preservation Group were essential to my understanding of the controversies surrounding the adoption of historic preservation legislation and practices, keeping me involved in design review meetings, and helping me contact homeowners to interview.

This research has benefited from many productive discussions with the following colleagues and friends: Kun Chen, Jack Fong, Luis Hoyos, Dolores Koenig, Gary McDonogh, Keith Murphy, Deborah Pellow, Bob Rotenberg, and Alan and Josie Smart. Their critical readings of text and thoughtful comments, in addition to those of an anonymous reviewer, have helped me sharpen my thinking and analysis. In addition, they have provided intellectual and moral support at stages, when my own enthusiasm for the subject lagged. A number of architecture students, including Robert Davidson, Jennifer Dowland, Victoria Huynh, Leslie Lum, Andrew Liu, and Jessie Waldmar, played key roles in data collection and analysis. I have been able to take advantage of numerous opportunities to present portions of this research at the American Anthropological Association and Society for Economic Anthropology meetings, and at the Urban Material Culture Initiative symposium sponsored by both DePaul University and the Field Museum in Chicago in 2011. I was also invited to present some of this material at Syracuse University, and at seminars in anthropology, architecture, and urban and regional planning departments. I am grateful for financial support in the form of a Cal Poly Pomona Faculty Research Grant.

Slightly different versions of some of the material appearing in this work have been published in earlier forms. The general thesis of the book appeared in a shortened version, in 2010, as "Cosmologies of Bungalow Preservation: Identity, Lifestyle, and Civic Virtue," *City and Society* 22 (2): 211–36. Material in Chapter 7 appeared, in 2014, as "Bungalows and Mansions: White suburbs, Immigrant Aspirations, and Aesthetic Governmentality," *Anthropological Quarterly* 87 (3): 819–54, and, in 2015, as "Residential Design Guidelines, Aesthetic Governmentality and Contested Notions of Southern California Suburban Places," *Economic Anthropology* 2: 120–44.

All the photographs were taken by the author, apart from those otherwise credited; the map was drawn by Nadim Itani.

Finally, I owe a debt of gratitude to friends and family members whose patience and support have played no small part in the completion of this project. To Judy and David Kronenfeld, and to Judith Sheine, I extend heartfelt thanks and appreciation for their long-term support. I am grateful to my son Scott for his enthusiasm for and curiosity about the topic of historic preservation that have sustained me. To Richard Zúñiga, my husband and best friend, go my thanks for accompanying me on this journey by attending countless historic home tours and related events, for his critical insights into the meaning of exclusion, and for his unflagging emotional and moral support for this project.

1

Framing Preservation

This study examines civic engagement over issues of neighborhood aesthetics by focusing on the power of domestic materiality. This power is situated and produced in the home as occupants decorate and remodel settings and their material contents imbuing them with significance. The concept of domestic materiality revolves around socially constructed practices that shape homeowner identity and lifestyles, but also extends into the sphere of civic engagement. Older suburban houses, especially those built over fifty years ago,[1] are subjected to physical alterations over the course of time, inside and out, as successive occupants use them for shelter, and to express their social status and aesthetic aspirations. Notwithstanding the fact that older houses are "used," or pre-owned, and affordable, this study seeks to understand why some southern California homeowners acquire them to renovate or restore. Those who study the ostensible agents of revitalization of older neighborhoods label these homeowners "gentrifiers," often dwelling on the disruptive effects of their efforts on existing communities (Brown-Saracino 2010). This study, however, shifts the focus toward understanding the underlying motivations and social contexts in which homeowners invest in the "historic preservation" of their properties and the ramifications of promoting aestheticized historic values in the public arena. The role aesthetic values play as they are embodied in public policies and aimed at controlling homeowner practices is the central question of this book.

Homeowners' deep interest in the materiality of their houses has the power to shape identities and lifestyles, but also to stimulate advocacy in the neighborhood and city for protections for "historic" resources such as individual houses, neighborhoods, and civic structures. The desire of homeowners to engage in civic activism and promote an aesthetic ideal of a residential landscape through legislative protections takes place in an environment where preservation professionals operate. The voices of these actors and other specialists can and do bring legitimacy to the arguments that intentionally exclude some architectural expressions while privileging others. Not surprisingly, these efforts may be resisted or contested in neighborhoods and at city hall by other homeowners who hold to a different vision.

Protecting Suburban America seeks to unravel the cultural logic and dynamics underlying the "preservation" of historic homes and neighborhoods in a selection of suburban southern California cities. The region is located along the foothills of the San Gabriel and San Bernardino Mountains, and was initially developed following statehood in 1850 for agricultural production, specifically citrus and fruit orchards. Communities were linked together by rail lines to facilitate commercial shipping, and to transcontinental rail lines connecting the region to the Midwest and East Coast. The rail lines brought migrants seeking better weather and economic opportunities in farming in the West. By the end of the nineteenth and early twentieth centuries, a sizeable population of gentleman farmers were engaged in citrus and fruit production, which they sometimes coupled with "retirement." Early real estate sales promoted 10- or 20-acre plots of land, some larger, all with irrigation rights. The first of these landowners constructed large Victorian houses, or grove houses, many of which still stand today. Later, as more migrants arrived, the landowners found real estate sales increasingly lucrative, and subdivided their orchards into smaller plots for the construction of individual homes, including smaller Victorian and Craftsman houses. Other properties were soon subdivided for the construction of very small bungalows and "kit-houses" to accommodate workers of various classes. Later, in the 1920s until the outbreak of World War II, land was continuously subdivided for the construction of speculative and custom-designed houses built in a variety of Euro–American styles as migrant demand for housing increased. Although the homes of the most wealthy and influential community members were usually designed and built by well-known local architects, designer–contractors and builders produced the majority of the houses serially, or in small clusters. In most cases, the investors, builders, and owners of these early suburban houses were ethnically white, while ethnic Chinese and Mexican workers were concentrated elsewhere.

The current state of these early twentieth-century suburban houses and neighborhoods constitutes the primary material culture around which this book is organized. From the 1950s until the present, homeowners have looked beyond just finding decent accommodation in a hospitable neighborhood: many also seek houses that are affordable, exude charm, and are close to work. As newer postwar housing was built, it was often located inconveniently far from business and civic centers. More centrally located older houses, however, often evidenced more wear and tear, and lacked the conveniences and modern amenities of the newer models. The more prior remodeling a house had endured, the more challenging the restoration of functionality and aesthetic charm; demolition can provide an easy way out. Thus, the materiality of the house plays a central role in new homeowners' efforts to realize their desires. An old house might constrain aspirational possibilities, but it also presents novel opportunities sometimes not initially imagined. As homeowners endeavor to produce their "historic ideal," they also encounter the agency of the materials they remodel. Through a variety of specific renovation practices, the material qualities of the house may "speak" to them as to the original intent of designers, builders or occupants. The biography of a house and its particular qualities exert a unique influence on homeowners, and shape through interactive processes the expression of idealized outcomes. The homeowners' initial encounter with the old house and their hands-on attempts to discern its physical composition reveal the first dimension of its material power.

New homeowners seek like-minded neighbors in a community to advise, support, and value their efforts in restoring and preserving a piece of the past. Since most

Figure 1 Traditional Craftsman house. Credit: Denise Lawrence-Zúñiga.

of the houses tend to be clustered in the oldest sections of cities, there are usually opportunities to bond with neighbors for mutual support and advocacy for municipal protections of "historic resources." Often their collective goal is to preserve or produce an "authentic" residential landscape by discouraging remodeling practices that are inconsistent with the original historic character. Municipal planners are keen to maintain property values and encourage revitalizing older districts by instituting legal protections under the guidance of preservation professionals.

Preservation-oriented homeowners typically collaborate with city officials in constructing specific local meanings and appropriate practices to ensure that residential and civic architecture accurately represents their understanding of their city's history. Collaboration and collusion also ensues as actors construct a common worldview, and adopt a cosmology that articulates a vision, rationale, and ethos for the material conservation of the most historical and original architectural elements in the suburban landscape. This second dimension of materiality's power is articulated in a preservation cosmology grounded in notions of authenticity, which authorizes the privileging of certain aesthetic expressions, and particular social groups, while ignoring or rejecting the values of others.

Restoring residential neighborhoods that have long suffered from disinvestment, or protecting homes threatened with modernizing remodels may be secured through historic preservation legislation, or other regulations enabling municipal authorities to control building design. Some homeowners' stylistic expressions may not be "histori- cally appropriate," for which preservation homeowners and professionals necessarily seek state intervention through municipal ordinances and policies to achieve a coherent aesthetic effect. The adoption of legal measures that encourage particular aesthetic qualities aim to motivate compliance by emphasizing homeowner design choices rather than forcing specific requirements. Design guidelines strive to teach and educate, rather than penalize, and typically fall under the rubric of neoliberal instruments of governmentality (Rose 1996). The specific techniques inevitably encounter resistance and opposition from homeowners who object to government intrusion, or whose ideal is incompatible with the city's official historic vision. Aesthetic norms promoted in city regulations by local officials, or more often through the subtle efforts of neighbors excludes some aesthetic preferences, while favoring others. At the same time, city regulations attempt to reform not just the design of the house form, but also the homeowner's vision of the ideal home in the idealized historic suburban landscape. The third dimension of domestic materiality's power is to use the individual house to reach into the neighborhood, to encourage and support some aesthetic preferences and social groups while excluding others.

The materiality of the house is both the product of social construction and the primary medium by which people shape their own identities and lifestyle practices. Materiality is not a side issue (Buchli 2013); it is central to suburban homeowners who adore older houses and want to see them restored to their former glory. It is also central to the way local governments operate, especially in the arena of land use and zoning, and in the somewhat controversial area of "community redevelopment," whereby quite large sums of money are made available, typically, to destroy the old and build anew. Municipal efforts to conserve "historic" sections of a city are increasingly popular in civic and commercial zones. Escaping under proverbial radar, however, is the fate of countless older and deteriorated residential neighborhoods that are also experiencing a revival due to newcomers' interests in aesthetic experimentations. Some remodeling experiments have been modest, such as covering bungalows with stucco, while other experiments, like replacing single-story houses with two-story "mansions" have been bolder. The suburban landscape of southern California is changing, and changing

Figure 2 Traditional Spanish Revival house. Credit: Denise Lawrence-Zúñiga.

rapidly. In the face of these challenges, this study considers the fate and future of many of these older suburban neighborhoods.

Protecting Suburban America begins with questions only homeowners themselves can answer. Given the many economic and physical challenges of owning and caring for an older house, why do people want to buy and renovate old houses? What is it that they see in an old house or bungalow that makes it irresistible and keeps them engaged in restoration over the long haul? How do restoration practices affect homeowner lifestyles and identities, and the way they see their neighborhood and city? What compels them to become active in education and advocacy through nonprofit organizations and at city hall? And, how does the institutionalization of preservation values and aesthetic preferences affect the future of these older suburbs? This study seeks to answer these questions by examining the experiences of homeowners, nonprofit preservation groups, and city officials in five foothill cities east of Los Angeles. The sections that follow review the literature that frames the inquiry, followed by a description of the research methodology employed.

The psychology and phenomenology of home

The changes in relationships that people establish with the physical qualities of their houses have been described in popular volumes and in scholarship that draws on a vast array of multidisciplinary research (Archer 2005; Bryson 2010; Rybczynski 1986). Architecture and architecture history, psychology, sociology, and anthropology, economics, and urban studies, to name a few, each discipline presents countless texts describing and explaining *the* definitive character of house and home. Aside from material characteristics that qualify them as shelter—possessing walls and a roof, at minimum, to provide protection from the elements—houses acquire myriad meanings and significance in the lives of people who dwell in them (Rapoport 1969; Oliver 1987). Houses are places made by people and are products of culture. Houses are useful for storing clothing, food and furnishings that make life comfortable. Residents inhabit houses, they dwell in them, and when they depart from the house, they regularly return to them. They provide the settings for the organization and social reproduction of family life and are imbued with meaning for staging and celebrating life-cycle events. By dwelling in the house, residents often impart something of themselves to make it their own (Birdwell-Pheasant and Lawrence-Zúñiga 1999). They may simply decorate the interior by hanging artwork or by painting the walls, or they may undertake a more substantial renovation by remodeling a kitchen or bathroom. Each act reconfigures the relationship the inhabitant has with the house, even if it is as simple as providing more convenience or aesthetic pleasure.

Psychologists conceptualize the home as the primary territory affording the individual the privacy necessary for forming self-identity and sense of self worth. Privacy, defined as "the selective control of access to the self" (Altman and Chemers 1980: 77), is considered essential for psychological boundary maintenance (Altman 1975: 49). Studies of spacing mechanisms among animals in captivity and the wild show territorial behaviors critical to survival strategies. In fact, anthropologist Edward

T. Hall (1966) used the territorial "fight or flight" concept in proposing "personal space" to describe culturally variable social distances as an unconscious form of non-verbal communication. People express territoriality by marking a space to control it and "defend" it if its boundaries are threatened or violated (Altman and Chemers 1980: 121–2). People often use territorial acts to achieve the control necessary for securing privacy, especially at home where privacy can be maximized (Altman and Chemers 1980: 129). Thus, while privacy's focus on "control" suggests protective practices in the service of the "self," territorial "marking" is essential in the construction of personal identity and maintenance of social systems.

Decorating a room or altering a house façade are territorial acts of personalization that express, or objectify (Miller 1987), an individual's or family's uniqueness, tastes, and values. These spatial acts employ symbolic elements to communicate personal boundaries, and imply control even if actual physical control is elusive. The association of décor with the "decorator" represents an extension or projection of the self beyond the body into physical space, which, if violated by trespass or alteration by another person, may provoke a defensive reaction (Altman and Chemers 1980: 137). People enmesh themselves in the material environment and acquire a sense of "ownership" of space through habitual occupation. Thus, a particular chair repeatedly occupied by a "father" soon becomes associated with him, even when he is absent. The father may feel quite attached to and protective of his chair, as if "owning" it, while other household members may defer to him and his authority by avoiding sitting there. Thus, household spaces and their contents come to signify gendered and generational status, power and authority, representing and reinforcing the domestic social order (Spain 1992).

While privacy, personal space, and territoriality imply an ever-present but inert and objective environment, the psychological concept of "affordance" suggests that the meaning of some material features depends on perceptions of them to "afford" or support certain human behaviors (Gibson 1979). A chair, for example, affords sitting while a door affords passing from one space to another. That is, the way that people read the significance and benefits of environmental features reside in their possibilities for their use (Gibson 1979: 127). Dant (2005) and Ingold (2000) maintain that affordance theory comes closest to a phenomenological understanding of an active and animated material order. Phenomenologists also try to understand relations between people and environments by stressing the lived experiences of spatial forms from a subjective point of view (Jackson 1995). Bachelard (1964 [1958]), for example, emphasized the role of an individual's memories in how domestic spaces are read and understood, and argued that the house we are born into "engraves" its functioning in us creating the "house as dream-memory" (1964: 15). These memories are critical to interpretations of the house, which have the capacity to reveal deep, unconscious socio-symbolic or cosmological meanings (Cooper 1974; Cooper-Marcus 1995; Douglas 1966; Eliade 1959; Korosec-Serfaty 1984; Levi Strauss 1963; Turner 1969; Van Gennep 1960).

Material culture

The material conditions of older homes play a dominant role in their rehabilitation and in shaping and re-shaping the suburban landscape. Discussions of home remodeling practices often focus on the mutually constituting qualities of social and physical environments and the role of the habitus, unconscious dispositions acquired early in life that guides the production and reproduction of cultural forms (Bourdieu 1977). As the child learns the repetitive and routinized knowledge of society's cultural traditions, that learning also leaves room for improvisation and change. Social life is a practical logic, but a logic also given to spontaneity, inventiveness, and imaginative creativity (Tilley 2006: 64). The habitus, then, does not completely determine or control, inhibit or impede new behaviors, but rather shapes individual agency in producing new, as well as routine outcomes. Thus, recollections of a childhood home and family life may be recalled positively, and so encourage the literal recreation of the material conditions evocative of those memories. Just as likely, however, is the creation of an entirely new domestic setting, based on selected ideas, gleaned from contemporary media or other experiences. Rehabilitating an older home invariably utilizes multiple strategies.

Home remodeling can render intangible phenomena like historic values and significance concrete and durable. The notion of "objectification" is used to describe the tangible embodiment of a concept, idea or value in material objects (Miller 1987, 1995a; Tilley 2006). Objectification refers to how people employ materials to produce an object that "becomes simultaneously a practice in the world and a form in which we construct our understandings of ourselves in the world" (Miller 1995a: 30). Thus, unlike territorial "marking," actors objectify and extend themselves in creating things, and, in that act, come to reflexively construct and know themselves. The object "speaks" to the homeowner about its proper material conditions, but also informs homeowner subjectivities. While subjects and objects are co-produced, ideas and values are made manifest, regardless of designer intentions, in the act of object creation, not before (Miller 1987). Moreover, material artifacts not only represent ideas and values, but they also reproduce and legitimize them. As homeowners select and incorporate these artifacts into their home remodeling repertoires, they pursue material integration or embodiment. The remodeled home environment becomes deeply integrated into the homeowner's sense of being there such that boundaries separating subject and object are no longer consciously recognized. Home and homeowner share an identity at the same time the material order affords a comfortable lifestyle.

The power of the material qualities of home, then, is found in actors becoming highly dependent on some of them in everyday life even while being mostly unaware of their dependency. These objects become "inconspicuous," or part of a taken-for-granted background, although they exert continuing influence by shaping behavior and expectations (Shove 2003). Objects are "naturalized," or are said to have "humility" (Miller 1987: 105), as they are incorporated into daily routines. But new objects present challenges for preservation homeowners who remodel old or restore modernized but deteriorated houses because they must be absorbed and normalized into the social and environmental fabric of home. A remodeling decision to substitute an early twentieth-century icebox for a modern refrigerator, for example, disrupts established networks that link it to other material objects, values, ideas, and routines. No longer are there any regular home deliveries of "ice" to fill the icebox, nor does contemporary life conform to this rather "inconvenient" technology. Access to newer (and older) technologies, such as the refrigerator, changes homeowners' expectations of appropriate cooking equipment, while also legitimizing their acquisition. Incorporating a contemporary refrigerator, or an icebox, and other kitchen technologies into a preservation-style kitchen requires some very creative solutions for function and aesthetics.

Similarly, the neighborhood landscape changes very slowly, it endures, and as such it has enormous influence largely because its presence is rarely noticed or examined. It is "background" and it is normal. "Naturalized" environments appear as "inevitable and timeless" and beyond question (Tilley 2006: 66), especially the historic residential neighborhood. Architectural artifacts have great appeal because they are not linguistic phenomena with discursive attributes, but are material with non-verbal and symbolic qualities (Tilley 2006: 62). Much of early childhood learning is pre-linguistic where presentational images are fixed on the unconscious mind through symbolic play (Miller 1987: 88); these images later serve to objectify feelings (Langer 1967). Our attraction to visual stimuli is strong precisely because it is emotional and defies rational explanation. The visual qualities of artifacts are ambiguous, they communicate multiple meanings, some of which may be contradictory and unconscious. To the extent that they speak, they often speak to and of unconscious sentiments and understandings, and their pre-linguistic visual qualities figure prominently in the construction of aesthetic values, meanings and preferences.

The powerful non-discursive appeal of domestic objects suggests to some scholars that they have agency, "the socio-culturally mediated capacity to act" (Ahern 2001: 110). Art works, for instance, may be perceived to move people, and are said to exhibit agency that invites inferences (or abduction) as to their causes and cognitive interpretations of their meanings (Gell 1998: 13). The artifact is both outcome and instrument of social agency. Primary agency is attributed to a person that creates an object; those material objects may be seen to possess secondary agency by people who believe the objects have influence over them. The primary agent, then, is the original source of the creative action, the authors or artists who make the object and distribute their own agency in the object's making (Gell 1998: 23). The secondary agency the object acquires is the result of the original maker's distributed agency, which enables the object to act independently of the creator's direct influence. That is, the power to cause others to act seems to lie within the object itself. As is often the case among preservation homeowners who seek to restore the original conditions of their house, that knowledge about the builder's or original owners' design intentions is critical to restoration, and that their personalities become known through the artifacts. Homeowners may also believe the material objects embody those intentions and speak to them about prior uses and proper restoration practices. They understand the objects to connect them to a lineage of creators and dwellers of their houses, while also placing them, the new owners, into the overall biography of the house.

An object's agency may be seen as acquiring power from the narratives or biography it acquires over its lifecycle. Objects have social lives through encounters with social and institutional systems, which re-contextualize them within different value regimes, especially those that commoditize them (Appadurai 1986). Once redefined in the marketplace as unique or common, object biographies may attach additional value considerations in the form of aesthetics, morality or religion (Kopytoff 1986). Historically preserved houses are especially susceptible to the use of these kinds of narratives to re-value their market prices. Again, object narratives can figure in individual life histories and the constructions of the self (Hoskins 1998, 2003, 2006), while collective narratives about public architecture and landscapes may attribute agency to large-scale artifacts. Many post-socialist European cities still contain buildings associated with unpleasant histories or have developed powerful negative reputations, often contested, for being "eyesores" that people reject (Van der Hoorn 2009). The eyesore label is often applied to older deteriorated houses and neighborhoods that have suffered from disinvestment before being rescued for restoration.

One of the most powerful dimensions of contemporary material culture is found in consumption practices, identity construction, and lifestyle (McCracken 1988; Miller 1995b). McCracken argues that domestic objects were used historically in Europe to verify a family's high social status and honorable reputation. Prior to the eighteenth-century, wealthy families of high standing used and displayed family heirlooms with signs of wear and tear or "patina" as symbolic evidence of their status claims (McCracken 1988: 32). Since such objects could be counterfeited, although not the patina, heirlooms offered a critical component for the visual verification of authentic status claims. By the eighteenth century, the rise of consumer culture began to erode the patina-status system since the wealthy found novelty a more compelling attribute for declaring superior social position. As subordinate classes quickly acquired similar objects, the wealthy were driven to acquire even newer and more fashionable possessions to advance their status claims. Although a new system based on seeking perpetual novelty has come to dominate contemporary consumption, the patina system still operates, especially in historic preservation and antique conservation. The restoration of older architecture depends on recognizing in durable material objects, once treated as "rubbish," the potential to be revived, recycled, restored or renewed to live again (Thompson 1979). Making the antique appear brand new, then, is not an indicator of taste, and possessing discerning judgment to "correctly" inform consumer choice may be far more important than actually possessing the object itself. These judgments are used as indicators of class status (McCracken 1988: 42; Bourdieu 1984).

Today's commodities link people through consumption to ideals and aspirations. The gap between what is "real" in everyday life and the "ideal" to which people aspire suggests the possible "displacement" of the ideal in historical time to a glorious past, or to a future time as utopic vision, or to a different place such as a colony or a colony's "mother country" (McCracken 1988: 106–7; Lowenthal 1985). Material objects can aid in the recovery of meaning, bridging the gap between real and ideal. A century-old bungalow restored to its original condition and filled with period antiques enables a homeowner to draw on the evocative power of antiques with which to create a bridge to the displaced meaning of a "simpler time" in history. Material objects are not just necessary for knowing who we are, but who we aspire to be (McCracken 1988: 117; Clarke 2001).

The appropriation of material objects in constructing identity and lifestyle is central to consumption theories that link individual actions to the larger sociocultural context. Buying and wearing the latest fashions is a marker of stylishness and affluence, but wearing pre-owned clothing could signal relative poverty as much as a stylish protest. Similarly, the style and décor of the house is capable of symbolizing and communicating the occupants' identity as traditional or modern, tasteful or tasteless. As a form of objectification the house is not simply a cover or shelter. Rather, the practices involved in living in the house, decorating it, or remodeling it are ways homeowners make the house their own and in the process make themselves (Miller 1995b: 285). The house imposes itself on decisions about how to decorate and live in the house, but it does so within the context of larger cultural trends. The "outdated" design features created by the original designer or builder, and that of subsequent occupants, present choices to the homeowner for salvage, restoration or demolition, and shape the remodeling decision-making process. Miller argues that there is something "intrinsic to certain properties of materiality" related to its durability and longevity that forces people to take positions "on wider cosmological issues of authenticity, truth and identity" (2010: 96). Thus, the durable features of the house may suggest to the homeowner "period décor" as a restoration strategy necessary for maintaining an authentic house.

One dimension of complex or large-scale home renovation projects is the reconstructive effect on homeowners' lifestyles. The deliberation over alternative ways of doing things, or different meanings, as well as seeking the integration of new practices and values into a pre-existing repertoire figure in reconstructing the homeowner's lifestyle (Lorenzen 2012: 97). These practices involve active engagement, self-reflection, and the construction of a new or evolving narrative of the self that explains the new coherence. At the core of the deliberative process is the consideration of the value of the history of their houses, but also how restoration practices can express a distinct "preservation" style and identity. Thus, the potency of material culture rests in its capacity as a sociocultural resource to organize people in the production of objects and houses that, in turn, produce meaningful ways for people to live. Homeowners are not only committed to uncovering the original designers' and builders' intentions and agency in their homes, but seek to infuse those material conditions with an historic imaginary by activating and "releasing" the original features and telling their story about it.

Community, cosmology, and authenticity (authority)

Although the key actors in home restoration practices are individual homeowners, they do not act in isolation; their restoration efforts require support from neighbors and municipal governments. Like-minded homeowners eager to engage in home restoration are drawn to neighborhoods in cities where older houses are concentrated. Two types of communities that share an interest in preservation are found and intersect in these settings (Cohen 1985). One is the place-based community that shares geography. This community includes owners of homes located in a geographically bounded area, but also craftsmen, professionals, and city officials who work there to preserve local history and culture. The other community shares interests in historic preservation without much regard for geography. It is defined by shared language, values and worldview, and particular knowledge about architectural conservation and preservation. It includes professionals, practitioners, researchers, and experts of various specialties including architecture and architecture history; urban planning and design; craftsmen, manufacturers, publishers, and media outlets; and local, national, and international nonprofit organizations. Membership in these two communities often overlaps and people from both work together, especially when advocating for preservation protections at particular sites, but individuals do not necessarily have exactly the same points of view.

The central concept organizing preservation communities gives primacy to the "original" material artifacts in constructing a cosmology that articulates meanings, values and a worldview. Cosmology, in this case a secular one, operates as a framework that organizes concepts, normative relationships and processes among phenomena that comprise "the universe as an ordered whole" (Tambiah 1985: 130; Herzfeld 1992). Although a rather elaborate and comprehensive cosmology of historic preservation and conservation is articulated for the professional pursuit of protecting monuments and civic buildings, homeowners and citizens interested in local history also become familiar with and interpret selected portions of the cosmology. As a secular cosmology, it presents a rationale for preserving the original building conditions, while also conferring legitimacy on the actions of advocates for legal protections. Local residents imitate the professionals, but their knowledge of laws, policy directives and research is often ad hoc, piecemeal and uneven. Homeowners pick up knowledge by participating in local historic preservation events, and from their experiences in renovating their own homes, while the formal training of professionals serves to effectively define and reproduce the cosmology, its underlying philosophical principles and practical applications. Nevertheless, both communities are similarly focused on discovering, recovering, conserving and restoring the original material qualities of the built environment, and they often join forces to advocate for policies to protect buildings.

To the extent that material features are believed, or verified, to be "original" and therefore evidence of history, they are taken to be "authentic." Rarely do preservation professionals and homeowners talk about rehabilitation in terms of authenticity, however, they do tend to obsess about the best methods for preserving original architectural conditions. On the other hand, the concern with presenting an "authentic" material representation of history has become a pressing issue in the heritage industry, and is problematized in modern tourism, heritage, and museum research (MacCannell 1976; Macdonald 1997). Authentic objects are generally defined as real, original, truthful, or genuine. Theodossopoulos, however, argues that there exist simultaneously diverse "sets of meanings that range from genuineness and originality to accuracy and truthfulness" (2013: 339). In the historic preservation profession the meaning of authenticity focuses more on the "original," in part because authorship or designer intent is primary. The authentic object must be trusted to be a genuine representation of the historic claims made to a less knowledgeable public. Architectural historians, beginning with the classical periods, are concerned with identifying names of individual building designers using *auctoritas* (Latin: authority, prestige) to refer to the "authority of its creator" (Favro 1989). In this sense, the authentic object carries the authority of its creator when the original conditions of its creation are known.

As an idea authenticity attracted special attention at the end of the Middle Ages as the feudal order dissolved, and the resulting social and material instability required stabilizing influences (Jones 2010: 186; Theodossopoulos 2013). Emergent scientific thinking constructed people and objects as bounded and discrete units with internal essences; an individual had an inner self while an object was composed of essential materials. The notion of a separate or inner self became even more pronounced after the Protestant Reformation, when an individual and their position in society could be known and verified by the person's "true self," conceived as distinct from all the roles an individual plays, while objects could be known by the purity of their unique essences (Jones 2010: 187). Authenticity, thus, became an important character-defining feature in the social construction of both people and objects.

Contemporary professionals who curate and display heritage objects to the public interpret the meaning of "authentic" that parallel concerns of preservation homeowners. The notion that authenticity lies beneath the surface or deep inside oneself is mirrored in homeowner practices that scrape off painted surfaces. Alternatively, authenticity is found in exotic societies "'uncontaminated' by modernity" and commoditization, just as an old house can present itself as untouched by modernity (Theodossopoulos 2013: 388). Heritage displays, however, are "cultural productions" and their staging may cast doubt on their authenticity and veracity (MacCannell 1976). The deliberate promotion of folk and vernacular architecture as authentic evidence of the cultural patrimony has served nation-states well in the creation of an imagined community (Hobsbawm and Ranger 1983; Anderson 2006). Ever wary of the unintended consequences of perceived falseness in staging, heritage conservation professionals rely on standards to represent history and culture and meet visitors' expectations for an authentic experience (Handler and Gable 1997; Jones 2010).

The professional heritage conservation approach has traditionally considered authenticity to be an objective and measurable attribute inherent in material objects (Jones 2010: 182). Conservationists promote techniques and processes to preserve the original material qualities in perpetuity, but constructivists argue that the authentic object has already been constructed as a cultural phenomenon and challenge these materialist or objectivist understandings (Gable and Handler 1996). Simply conserving an artifact requires physical alterations to remove previous modifications, which can undermine the artifact's "authenticity." Moreover, constructivists note that material objects alone cannot convey all the historical meanings necessary to their understanding. They require a narrative account to frame and interpret the object so that it is believable, and, through transparent staging, retains its reputation as authentic (Gable et al. 1992; Bruner 1994; Gable and Handler 1996: 578).

Contrasting notions of materialist and constructivist conservation approaches inform how historic preservation cosmology constructs meanings, codifies practices, and differentiates between professionals and ordinary homeowners. Protective legislation for historic architectural resources in North America draws heavily on strategies for the conservation of archeological materials. Today's regulations and procedures outline authentication requirements, including identifying the object's historical "significance" and its integrity, or capacity of the object to materially represent that history. Guidelines also prescribe proper strategies and techniques to help maintain and rehabilitate or restore the object's original features (Tyler 2000). Some prescriptions are highly technical and require particular knowledge and skill that only preservation specialists have. The extent to which strict preservation standards are demanded and applied varies considerably, and depends on the national, state or local historical significance of the object, and the locale.

In restoring original architectural features, ordinary homeowners may employ less than orthodox, more "flexible" techniques. Sometimes they receive corrective advice on strategies for rehabilitating material features when visiting city hall for permits. The aim of homeowner conservation, however, does not necessarily include considerations of authenticity before a critical public or other professionals, but is confined to the homeowners' private enjoyment of the house. In fact, formal legal preservation protections usually apply only to the exterior of the house. Cities may grant homeowners some latitude in interpreting preservation standards for house renovations, but homeowners who serve on local commissions may also interpret restoration requirements more stringently than preservation professionals. Inspired by preservation cosmology, homeowners may adopt arguments for the authority of the original material conditions to justify their remodeling decisions, and to legitimize campaigns for local historic preservation laws. That is, the homeowners expand the sphere in which to apply their moral argument to include their fellow citizens who may not agree with the premise of restoration. Thus, professionals and homeowners understand authenticity of the house differently, with homeowners emphasizing a constructivist approach based on the experiential, and less rigid interpretation of material conservation requirements.

Homeowners' restoration interests emphasize discovery and construction of the self over the production of an accurate material representation of historical facts. Many learn a variety of professional preservation and restoration techniques, but also draw on archival research and social contacts to construct an explanatory narrative. Their inquiries often lead them to discover the original designer's intent, and that of later remodelers, and match that knowledge with the material artifacts they uncover in their renovations. They understand their house to have a unique history and relationship with a network of past and present actors that give it its own agency. The material features homeowners uncover embody social attributes that lend an aura or "voicefullness" to the house, and allows homeowners "to achieve a form of magical communion" as they become personally incorporated into the network (Jones 2010: 189). The role of material culture in home restoration work is often personal and subjective. In this study, the particular agency of current and past inhabitants is found united in negotiations with the materiality of the house as new homeowners construct and complete their own stories and selves. When historic artifacts are considered "inalienable possessions" of the house, its authenticity is uniquely defined by "all the past experiences, people and places with which they have been connected" (Jones 2010: 190; MacDonald 1997).

Although preservation homeowners appear to engage in restoration practices in isolation, in actuality their activities take place in the extended social and moral context of the suburban neighborhood, where preservation sentiments are promoted. Like lifestyle migrants who search for a more authentic way of life in the countryside or in another country (Bruner 2005; Hoey 2014), preservation homeowners may pursue "authenticity as a social process" (Benson 2013: 519; Brown-Saracino 2010). The performance of the homeowner's preservationist identity within a community of enthusiasts distinguishes one homeowner from others in the neighborhood, a role evidenced by the homeowner's critical knowledge of material culture, by "taste," and in the physical appearance of their home. Given the economic and logistical challenges some homeowners face in actually acquiring "original" material objects, the preservationist identity rewards homeowners for creatively elaborating their taste and historical knowledge. Performing the preservation identity is meant to reveal discerning judgment about the architectural features idealized as original and authentic in the preservation cosmology as a form of entrée into an exclusive group.

Social inclusion and exclusion in the suburb

The performance of taste is about social inclusion and exclusion. It is also about place making. Preservation homeowners have not been shy about using arguments grounded in preservation cosmology and ostensible concerns for local history to advocate for legal protections in their communities. In many cases, the protections are sought in response to a perceived threat from a particular group such as developers or other residents who do not share their worldview. Duncan and Duncan (2001, 2004) argue that landscape taste is a form of cultural capital and that the aesthetic production of landscape as a nature preserve acts as a mechanism for exclusion in the New York town of Bedford. There, zoning regulations and environmental protections implemented in the "quest for social distinction" make place exclusive, with an aura of scarcity, and therefore a positional good. Seeking historic preservation legislation to protect residential structures or entire districts acts in a similar way. Protective regulations establish exclusionary zones that, under the guise of historic significance, create new boundaries, naturalize the neighborhood landscape, raise property values and engender a sense of residents' entitlement (Duncan and Duncan 2001: 390). Although protective measures focus on the normative aesthetics of residential places, in practice they have deeper exclusionary implications. To prohibit a homeowner from remodeling their property according to their own tastes, while requiring them to meet municipal design or historic preservation standards, is an exclusionary act.

The role of local government is critical to preservation advocates for securing protections for what might be seen as their own aesthetic preferences. Homeowners try persuading others through public education about local history, or organize educational tours and demonstrations, but without some kind of government intervention their aesthetic aspirations are unlikely to be fulfilled. For municipalities to privilege preservationists' aesthetic values over others is often rationalized as a civic duty meant to acknowledge local history, even while history is not a health or safety issue. The city's image or "brand," and concern for its own reputation may be deemed reason enough, but achieving compliance takes more than educational outreach. In the world of contemporary urban governance, neoliberal notions that couple market solutions with individual freedom to make choices help reframe the issue. A "new formula of rule" depends less on direct government mandates and more on using experts from a "market governed by the rationalities of competition, accountability and consumer demand" to devise policies (Rose 1996: 147). Governing is conducted by regulating the choices presented to individual citizens who are assumed to want to satisfy aspirational goals through consumption. At the center of neoliberal governance is the idea of governmentality, or government rationalities, which emphasizes how individuals in their communities learn from policies and other codes of conduct to govern themselves (Maskovsky 2006: 76).

Adopting historic preservation legislation requires municipalities to institute procedures such as design review and publish style guidelines for house construction and remodeling. These measures constitute the key "educational" vehicles for presenting residents with appropriate aesthetic and design choices. As a form of "aesthetic governmentality," these policies impose a dominant aesthetic vision of the city and reconfigure residents' "sense of self and place" to conform to an ideal (Ghertner 2010: 207). Design guidelines, using photographs and drawings, propose an alternative aesthetic reality, and descriptive "narratives " bolster the imaginary spatial experience, creating expectations of how a space "should" look (Ghertner 2010: 208; Legg 2005: 148). Although visual media are emphasized in changing subjectivities, the process also requires an official hearing or design review ritual to ensure the messages are conveyed and reinforced. Although petitioning homeowners present their own aesthetic preferences, which may not conform to the expectations of the city or preservationists, design review provides an opportunity to negotiate and win approvals.

Promoting aesthetic norms through preservation programs is embraced increasingly as a neoliberal instrument of civic and economic revitalization, especially in urban areas suffering from long-term disinvestment (Smith 2002; Wyly and Hammel 2005). Historic preservation is a "complex cultural project" aimed at realizing the imaginary vision, the "preservationscape," is often a key tool in urban redevelopment schemes (Wilson 2004: 43, 51). As a rule, urban renewal is associated with the creative class of artists and architects who turn dilapidated buildings into trendy settings, boosting real estate values and popular appeal, and restructuring the city (Zukin 1989; Jessop 2002; Florida 2010). Central city neighborhoods with lower income, often minority, residents are targeted for their attractive real estate values, and, as they are transformed, the creative class displaces the original inhabitants and gentrifies the area (Smith 2002). On a global scale, established European urban centers reclaim working-class neighborhoods (Herzfeld 2009), while in developing countries such as India and China, impoverished slums, or deteriorated historic neighborhoods, are targeted for demolition and renewal (Ghertner 2010; Arkaraprasertkul 2012). It is not clear if these latter cases, in fact, are also examples of gentrification as defined in the West (Ghertner 2014, 2015).

The pursuit of historic preservation in the suburbs often lacks the urgency of urban economic renewal and, while it may be given support and protections by sympathetic civic leaders when homeowners later demand it, that support does not always act as the primary stimulus. Rather, market forces or "back-to-the-city" consumption sentiments are more likely to inspire middle-class homeowners to discover older neighborhoods. In the United States and western Europe there may be three or more waves of "spontaneous" gentrification that began in the early 1970s and continue to evolve (Hackworth and Smith 2001). As new homeowners rehabilitate the old houses, they create conditions to attract others, which over time leads to the institution- alization of businesses and civic support, and establishes a new identity for the area (Clay 2010 [1979]).

Although conflicts between newly arrived homeowners and long-settled residents are well known (Williams 1988), gentrification's slower but more effective action is in the displacement of lower- and working-class residents that, on the one hand results in a rise in property values for homeowners, but also raises rents and forces renters to seek new lodgings (Brown-Saracino 2010: 3; Smith 1996, 2010). Gentrification, however, is not the inevitable result of historic preservation strategies, since it can also be used to ensure that elites continue to control neighborhood aesthetics and social groups. Restructuring the suburbs by class, and often ethnicity, also remakes the image of the historically preserved neighborhood. The regulations and practices aimed at restoring or maintaining the original Euro–American architectural styles also promise to reproduce the older "white" suburb, even though this time the remaking is largely symbolic.

Historically, the most exclusionary notions about the American suburb are grounded in the tacit understanding of its whiteness.[2] Originally settled by white European– Americans from the Midwest and northeast, older southern California suburban communities were built using idealized Euro–American housing styles. From the end of the nineteenth century until around the 1920s, these communities were inhabited by white, mostly middle-class farmers and businessmen, with ethnic minorities such as Mexicans and Chinese segregated to their own communities. Institutional practices aimed at excluding nonwhites and preserving the ethnic concentration of whites in the suburbs included the incorporation of restrictive covenants in deeds that prevented the resale of houses to nonwhites, and FHA (Federal Housing Administration) and bank redlining practices[3] that restricted lending primarily to white homeowners (Hayden 2003; Harris 2006). Lipsitz (1995) has argued that these legal systems and public policies promoted a "possessive investment in whiteness" that not only excluded some ethnicities but promoted the homogenization of European–Americans as diverse as Italian, Polish, and Irish, making them unified and "white" in the suburbs. After World War II, the federal government invested heavily in new suburbs where whites were encouraged to settle as opposed to the aging neighborhoods in inner cities where minority populations had taken up residence.

The material qualities of contemporary suburban neighborhoods are critical to the production of suburban whiteness. Even though earlier discriminatory policies were eventually rolled back, older minority neighborhoods continued to experience both deterioration and modernization, which further eroded potential historic value. Historic preservation policies, which are most often promoted by the middle classes, may displace working-class residents in these neighborhoods, but ethnicity also plays a role as a covert category underlying suburban change. Ethnicity is not an explicit exclusionary issue in any of the cities considered in this book, but taste and aesthetic preferences often are. Quite often, however, class and ethnicity are markers for disapproval. Middle-class residents also differ amongst themselves about aesthetic values, whether historic or modernist designs are preferred, but preservation homeowners as a group include households of diverse ethnicities. What is meant by "whiteness" in the evolving historic suburban neighborhood, then, is not the concentration of ethnically white people, but the concentration of "white" taste, "white" historical values, and "white" culture to which anyone can ostensibly aspire. The consequence of restoring older homes to their original conditions produces the idealized "white" suburb that recalls the original settlers and their time, but also produces a counter aesthetic to the sameness of new tract developments, the chaos of uncontrolled remodeling, or continuing decay.

Southern California provides a fairly welcoming melting pot for all kinds of relatively affluent immigrant groups from Asians to Latinos, and Armenians to Persians, each expressing inclinations for distinctive house styles and their own culturally informed aesthetic preferences. Their arrival in older neighborhoods, however, poses one of the newest threats to historic preservation values and local advocates' efforts: the introduction of new mansion-sized houses. Homeowners are required to pass through a city's historic preservation or design review to get a permit for remodeling, a requirement Ong (1996) likens to a "citizenship test." The design review, a form of aesthetic governmentality, requires applicants to focus on aesthetic qualities and, often, local traditions, which many immigrants perceive as subjective or simply do not have. The burden of exhibiting "white" taste and values rests with the applicant, which effectively screens out and excludes alternative aesthetic and cultural expressions. Thus, in the absence of restrictive covenants or other overtly discriminatory policies, municipal preservation procedures operate as techniques to facilitate self-policing and self-censoring in the production of the suburb. In the same way that historic preservation protections result in the exclusion of lower-income homeowners as a matter of class-based aesthetic values, they may also exclude middle-class immigrant homeowners whose aesthetic preferences do not reflect the "whiteness" of the suburban ideal.

Global gentrification

The changes felt in southern California's older suburbs parallel those experienced in many communities in the United States that are protected by historic preservation laws and policies. Large-scale urban revitalization projects or the preservation of large stately homes in major historic districts are quite different from the rehabilitation of suburban homes and districts. Sometimes preservation policies are initiated, embedded by the municipality into urban revitalization schemes in order to conserve its cultural resources. Other times new homeowners move into an older suburban neighborhood, perhaps displacing existing inhabitants in the process while demanding historic preservation protections from the municipality. In some suburban neighborhoods historic preservation protections may provide residents a strategy, for exclusivity's sake, to keep out those whose aesthetic values are perceived to be different. In other cases, preservation-inclined homeowners engage in a battle of aesthetic preferences in home remodeling with other homeowners who are well funded but prefer a different aesthetic. Although some of these examples may be labeled "gentrification," primarily because inhabitants are displaced while new homeowners invest capital in the property rehabilitation, not all situations where historic preservation is invoked fall automatically under that label. Rather, the use of aesthetic guidelines as an exclusionary principle can have different outcomes in practice.

According to some international scholars (Ghertner 2014; Smart and Smart n.d.), the use of the term "gentrification" may have been broadened inappropriately from its original meaning to include examples of population displacement in other parts of the world. Some of this displacement is associated with the demolition of slums to make way for the construction of modern, high-rise apartment buildings, or historic preservation efforts concentrated in and around major monuments and architecturally significant structures meant to attract tourists. These state-organized development schemes acquire land through eminent domain or other means, while often failing to provide for the displaced original residents. They intentionally "gentrify" the city by introducing younger, more affluent homeowners who are eager to shop and use public spaces and amenities. Like the historically preserved neighborhoods traditionally described in the United States and Europe, these new settings are described under the theoretical framework of gentrification. The marginalization and exclusion of lower-income residents occupying these sites, however, are not necessarily accomplished through voluntary movement, but by state force (Atkinson and Bridge 2005). And, not all settings described as undergoing gentrification are the product of the same kinds of economic, political, land use or aesthetic pressures. What many of these settings share in common, however, is the desire for an aestheticized vision to selectively exclude some residents while privileging the presence of others.

Methodology

This study examines the preservation of early twentieth-century houses in five southern California suburban cities located in the foothills of the San Gabriel and Pomona Valleys, east of Los Angeles. By the end of the first decade of the twentieth century, these cities, including Pasadena, Monrovia, Alhambra, Ontario, and Riverside, had incorporated.

Initially they were settled as agricultural "colonies" during the late nineteenth century, and specialized in citrus production, innovative irrigation technologies, and the introduction of railroad transport for passengers and products (Dumke 1943; Kershner 1953). Agricultural production was soon followed by real estate speculation with farmers subdividing and selling sections of orchards to eager migrants from the Midwest and East coast (Gentilcore 1960). Many of the earliest Victorian-styled houses were built and occupied by these earliest affluent farmers, but by the twentieth century Craftsman-style houses and bungalows became popular among

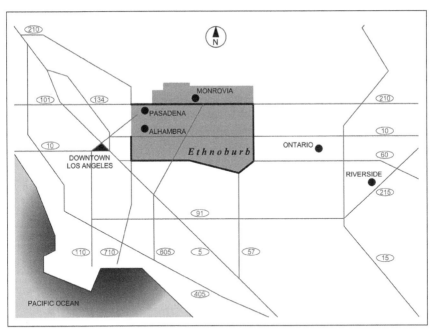

Figure 3 Map of southern California region indicating five cities and the Asian settlement area, the "ethnoburb." Courtesy of Nadim Itani.

upper-middle and middle-income residents. As civic and commercial zones were filled in, housing construction also shifted to a variety of "revival" styles, such as Spanish or Mediterranean, and Tudor. Interest in the preservation of historic civic buildings did not arise until after World War II and was often precipitated by the demolition of a landmark building, a Carnegie Library or theater. These events stimulated the organization of civic historic preservation groups in some local communities, and, with the state's adoption of preservation legislation in the 1960s, created a corps of experts and specialists to work within and outside the local governmental structures. As the preservation movement established itself and spread to many local communities, it also produced many businesses of design consultants and architects, contractors and craftsmen, and stores that specialize in products for the older house.

The primary modes of data collection included in-depth, semi-structured interviews for open-ended interviews and participant observation, including attending meetings and events, in each of the five cities. Photography and archival research from official city documents and news sources provided additional sources of data. Interviewees included five to seven homeowners in each of the five cities, city planners who are involved in historic preservation work, and leaders in civic preservation or advocacy groups. As many "historical" figures were interviewed as possible; these included founders of civic groups and former city officials. I should note that as an anthropologist who is also a professor in an architecture department, I had access to colleagues in the departments of architecture and urban and regional planning who are personally and/or professionally involved in historic preservation. These colleagues have been extremely helpful in referring and introducing me to key players in these cities and other professionals.

Homeowner interviews were primarily conducted with people who had bought a house with the intention of remodeling it to its original conditions, or who adopted a restoration strategy by living in the house. Other homeowners who opposed or resisted historic preservation influences, or who had owned their houses for many years only to discover preservation interests recently, were also interviewed. Questions focused on discovery and restoration strategies and permitted subjects to expand into narratives of their preservation strategies. Probes also encouraged subjects to reflexively discuss their learning process as well as their future aspirations and past accomplishments. Homeowner participation in neighborhood or citywide organizations and groups was queried, including specialized roles and responsibilities, and perceptions about neighbors and other citizens about preservation. Finally, homeowners were asked their perceptions about city government's role in historic preservation in their neighborhood and the city. These interviews, which were digitally recorded, ranged from one to four hours, they were conducted in the interviewee's home, and were usually accompanied by a guided tour of the homeowner's house. The tours were critically important in facilitating an understanding of how homeowners saw the remodeling challenges and the solutions adopted for solving them. A copy of the interview protocol is included in the appendix.

In addition to homeowners, city planners were interviewed. These interviews focused on the official roles and responsibilities of planners who oversee the preservation program in their city and the challenges they face. Questions probed for issues with regard to the legislation and review process, including public understanding and compliance, the division of labor for reviewing applications or dealing with violations, and relations with local civic groups, other city departments and the city council. In addition, members or officers of neighborhood or citywide preservation groups were interviewed. Most questions focused on organizational challenges and relations with the city. A copy of this interview protocol is also included in the appendix.

In four of the five cities, subjects for homeowner interviews were secured through preservation organizations; often they were active members in the organizations themselves. This selection process undoubtedly introduced some bias in the sample by introducing me to people who were perhaps more enthusiastic about preservation than their neighbors. It is clearly the case that not every preservationist homeowner in a neighborhood or city belongs to the organization or participates in civic advocacy activities. And it was known that some very committed preservation homeowners wanted nothing to do with the local organization. On the other hand, some organization members were more passionate about history than its material manifestations. Even so, the sample in each city produced enough variety of engagement and passion for preservation to ensure a range of values and practices. When it was possible, I also attended board meetings of the neighborhood group or civic organization. In two cities I became a member of the organizations in order to receive newsletters and announcements of meetings. I also attended historic preservation or cultural heritage meetings in each city and municipal design review meetings when issues affecting the particular neighborhood I was studying were addressed. And, I attended home tours in each city and participated in other social events whenever possible.

There are obviously more than these five cities in the southern California area from which to sample. At the outset, no objectively constructed selection strategy was employed, although I avoided some "exemplary" cities recommended by preservationist colleagues who wanted to promote their favorite preservation program that had high standards. This sample was then comprised, somewhat opportunistically, of cities with which I had had earlier personal or professional contact, but the main goal was to select sites characterized by differently structured preservation protections. There certainly could be more cities with other differences added, but in an attempt to limit the pool, I stopped at five. I wanted to know, for example, if cities that had many historic districts had different experiences with preservation than those that do not. I also wanted to know whether the presence of a neighborhood or district preservation organization made any difference in the experience of residents. And I wanted to know what difference it made if there was no historic preservation legislation in a city where advocates lobbied for its adoption. Although I live in the city of Pasadena, the neighborhood I studied there is not my own.[4] In Pasadena, residents can petition the city to establish an historic district, and there are many. Monrovia has only one small district and favors land marking individual houses instead. Both of these cities have strong civic organizations, and in Pasadena the district and citywide organizations are both very active. Riverside and Ontario, on the other hand, allow for residential districts and also have citywide organizations, but these are not necessarily very active in particular neighborhoods and, in the case of Ontario, the citywide organization seems virtually inactive. Here, the city planning departments are quite active in protecting residential historic resources the residents seem less enthusiastic about. Finally, in Alhambra where no historic preservation legislation exists, an active group of preservation advocates wage a continuing battle to get the city council to adopt stricter protections than the residential design guidelines they approved in 2009.

In the chapters that follow, the role of domestic materiality in the construction identity, lifestyle, and civic activism is explored in the lives of historic preservationists in the five cities. Chapters 2 and 3 explore the intimate details of discovering, uncovering and recovering the "original" features of the old houses homeowners have bought. Much of the discovery is associated with identifying the original designers or owners, and the intentions they expressed in material form. These discoveries give rise to sentiments of material agency as the personalities of previous actors become known and enmeshed with the current owners' lives. In Chapter 3, homeowners talk about a variety of strategies for restoring the original conditions of the house, including "purist" approaches, and finding fulfillment in the kind of identity and lifestyle the restored house enables. Chapter 4 describes the cosmology of the historic preservation movement that emphasizes material conditions and underlies legislation and scholarship, and guides professionals at city hall in applying its standards. Chapters 5 and 6 describe the histories of adopting preservation protections in each of four cities where the structural differences reveal different sociocultural dynamics and discourses over contested landscapes. In two cities, with elite families of long-standing, which have come to embrace preservation protections as an explicitly exclusionary device, issues dealing with local organizations are subdued, and city reviews are handled at a distance. In two other cities, in which a newer population of homeowners has gentrified run-down areas, activism is strong and participation in city reviews is quite active. In Chapter 7, Alhambra is discussed as ground zero for the advocacy of historic preservation legislation. Alhambra's demographic makeup has changed significantly since the 1980s, as more Asian, primarily transnational Chinese residents have moved in and have sought to remodel their older houses. These demographic changes are not limited to Alhambra alone, but affect the entire San Gabriel region. Alhambra is also home to active preservationists who oppose regional tendencies to "mansionize" and modernize old houses preferred by the new immigrants, making design review an arena of highly contested discourse over the aesthetics of residential landscapes. In concluding, Chapter 8 considers the five southern California cities comparatively, and sets the study within the context of North American and international historic preservation and conservation practices and gentrification.

Discovering Material Agency: Making the Preservation Homeowner

Given the obvious challenges, what makes people want to buy and renovate an old house? To reverse the deterioration that comes with age, or undo the remodeling mistakes made by previous occupants, seems to reinforce the classic image of the old house as a "money pit." Renovating a house with "sweat equity" typically requires a homeowner to invest long hours and physically demanding effort on weekends and vacations, not to mention how it inconveniences everyday living. These demands make the task of restoring a house to habitability a steep uphill climb even for the most committed. Although not every homeowner necessarily seeks to return their house to its former glory, many who have done it laugh when asked why. They launch into stories that suggest initially underestimating the amount of work required, but also tell of the joys and mysteries, the rewards and satisfaction that such a "struggle" entails. The purchase of a house is said to be one of the largest single investments Americans make during their lifetimes, but the purchase of an old house, especially when one could afford something newer, suggests a motivation more powerful than the desire to meet practical, utilitarian needs for shelter. The acquisition of an old house requires consideration of the complex aspirations homeowners have for their new purchase.

This chapter examines the narratives homeowners construct to explain their initial attraction to the old houses they buy, and to describe the conditions they find and gradually uncover as they began to live in them. As many scholars have pointed out, the house is central to our sense of who we are; it is a space in which we achieve privacy, find refuge, express ourselves, and construct our identities (Archer 2005). It is both symbol of the self and a laboratory for self-realization (Cooper 1974). The construction of identity, however, is not a result of passively occupying domestic space, but the product of ongoing interactions inhabitants have with their physical surroundings. The idea that we are molded from an early age at home by our physical as well as social surroundings is captured in Bourdieu's notion of the habitus, the unconscious dispositions that shape our inclinations to see and act in the world (1977). For the buyer of an old house, the habitus plays a key role in stimulating attraction, influencing expectations, and driving aspirations. Of course, not everyone who buys an older house is necessarily interested in preserving or restoring it, even though they may later develop an appreciation for a home's antique features. In fact, many people buy older houses because they are affordable, and they dream of modernizing and upgrading conditions without regard to salvaging original features.

Among the thirty-six homeowner interviews considered here, all have preserved or expressed an interest in preserving or restoring at least some of the original features of their houses, but the degree of commitment to preservation varies in intensity and quality. Interviews conducted in the homes of three elderly homeowners, who were couples over the age of seventy-five, for example, found that they had lived in the same houses for over forty years. None of them purchased their houses with the intention of restoring original features, but they had come to appreciate the special qualities their houses possess as they and the houses aged together. Another three interviews were conducted with homeowners who have a personal or family connection to the house—either they grew up in it, or one of their ancestors had built it. While one might assume that the relationship family members establish with their ancestral homes would be self-evident, in fact, their relationships are as complicated and varied as individual biographies can be. Thus, family-connected homeowners are worth considering separately at the end of this chapter to reveal these differences.

For homebuyers who become avid preservationists, dispositions of the habitus play an important role shaping the investment of energies in transforming their homes and themselves in the process. First encounters with an old house sometimes trigger nostalgic memories or its specific material qualities captivate and charm, and the initial occupation may bring pleasing surprises, or shock and horror. Homeowners' sustained interest in restoring an old house derives from the social relationship they form with the material qualities of the house itself. They often make their restoration into a moral project about how occupants "should" live in their houses. The house comes to play a central role in family life and serves as a vehicle for discovering the lives of prior owners and occupants. Restoration inevitably involves uncovering original material features, the agency of which may be revealed in homeowners' narrative accounts of their early encounters with the house. At first, the house reminded them, moved them, communicated with them, or made them feel certain emotions that encouraged them to acquire the property or stimulated their restoration efforts. Some homeowners spoke of their houses as having a voice or desire, attributing human qualities to inanimate objects. Family-connected homeowners have particularly rich stories to tell because their relationships with the houses are so much more personal than those who work to construct a connection with previous occupants. And, while elderly homeowners, who have lived many years in their houses, are less likely to attribute agency to the house as part of their initial attraction, they also speak with affection about its material qualities as evocative and memorable settings for fully living family life. In all, homeowners established meaningful relationships with the materiality of their homes by living in them, exploring their histories, and maintaining and remodeling them, and in the process were transformed by them.

The concept of material agency is central to the narratives told by preservation homeowners because it focuses attention on the perception of physical objects and the development of personal attachments and social relationships with houses. Although this chapter is primarily concerned with homeowners' initial encounter with their houses and the hint of agency, Chapter 3 explores remodeling practices that bring about a deepening awareness of the house's agency, the formation of an intimate relationship, and the reciprocal effects on homeowner identity and lifestyle. Describing material objects as if they had agency or human will may seem somewhat unusual, but anthropologists have long speculated about the role of material objects in exotic societies and our own. As gift or trade items circulate and are exchanged, people attach stories to them, and then the objects themselves become celebrated because of their association with celebrated "owners." Over time the objects seem to acquire biographies that enable them to have influence independent of specific human actions (Munn 1986; Weiner 1992; Hoskins 1998, 2006; Kopytoff 1986).

As physical objects houses do not, of course, actually travel or circulate. Rather, designers, builders, owners and occupants pass through houses and any celebrated personalities among them may impart some of their renown. Homeowner narratives, which focus on the reputation of the house in the community, cite historic associations with founders of the city, early entrepreneurs, wealthy landowners or other named celebrities of local history. It is not unusual for homeowners actively to construct or re-construct house biographies. These celebratory associations imply value, and may transform an old house into a recognized "historic" or landmark house worthy of protection. Thus, the celebrated houses have the capacity to influence the behavior of their occupants and others by engendering respect and, presumably, care. The house acquires a personality and identity and, while buyers may know something of the history in advance, more often they discover and reproduce it after moving in. The hint of agency encourages those buyers who are so inclined to learn more, and that can only be achieved by deliberately establishing a personal relationship with the house through intentional remodeling practices.

Sometimes the reputation of the old house is associated with a named architect or designer,[1] such as the Greene Brothers, who achieved national recognition for their Craftsman-style houses in southern California, or a locally recognized builder or carpenter. Architectural objects are artistic works that acquire powerful meanings as a result of how architects impart some of their own agency or personhood into the objects that they create (Gell 1998). By transferring their own vision and energy into designing and making the art object, artists expect the object to independently communicate with the viewer. The object possesses a kind of secondary agency derived from the artist's own communicative intentions. Although architectural designers and craftsmen work with artists' expectations, the house is also habitable space. It is meant to be used. It is not just an object for contemplation, although it can act that way, but must also meet human needs for shelter, comfort, and refuge. As inhabited space, houses are embodied in ways unlike other architectural forms; occupants' knowledge of them is intimate and is grounded in the reciprocal relations established through daily living.

Most old houses were neither owned by a city founder, nor crafted by a celebrated designer; they are just old and ordinary, even mass-produced, and, in the past, were owned by average citizens, lacking stellar biographies. The house is more apt to be experienced as having independent agency than other objects, in large part because the house endures longer than we do and is, therefore, capable of exerting influence over us (Miller 2010: 95). This view suggests that because a house is old, it has managed to survive, and that survival alone imparts special significance that compels our attention. At least it has the capacity to reveal something to new owners about what it has endured over the years. Initially, the house presents itself to the receptive homebuyer as a mystery—waiting to have its secrets uncovered and exposed. The homeowner who undertakes restoration, then, is confronted with quaint or curious material features better suited to an earlier age, or they are anomalous or deteriorated. To some homeowners the features seem to beg for understanding before they are unceremoniously thrown out, so as to be carefully restored. Access to photographic, written, or oral documents is helpful in establishing original conditions, but homeowners are often left to make sense of their houses by personally investigating their material qualities, conducting comparative research and making inferences from a handful of material clues.

The first encounter …

Although the house may be old, especially when compared with contemporary housing stock, its design features often appear novel on the first visit. Purchasing an older home may be motivated by the initial encounter triggering fond memories, nostalgia, or the appeal of well-crafted antique qualities. Homeowners often recall the first visit to the house as making a strong impression, and they weave stories about features that initially attracted them. The specific material features that trigger a positive response and stimulate a desire to acquire a particular house may be associational or ontological. The style of the house or particular features sometimes reminds them of a favorite relative or childhood experience, or the craftsmanship of its antique qualities holds their fascination. All homeowners do recall some practical reasons for purchasing their houses including the size, location, or price, but those requirements may only set parameters within which other criteria are used to choose.

The three elderly homeowners who had bought their houses at least four decades earlier, before the houses were considered old enough to be worth preserving, for example, said they were mainly attracted to the size and affordability to accommodate their growing families. In 1968, when seventy-seven-year-old Marvin bought his 1922 Monrovia house for his growing family, his own father, who helped him with home repairs, said that he thought the house was "modern." Although it may not have been "modern" by postwar standards, Marvin says that he didn't think of it as a historic house: "I don't think I thought about it that way." By contrast, younger homebuyers who become passionate promoters of preservation most often mention an initial encounter that peaks their curiosity and "draws them in." Even the three family-connected homeowners mention specific features or settings that they had shared with now departed family members as triggering their desire to go back. Thus, homeowners' nostalgia about memories of family members, or first encounters with old houses set the stage for anchoring and explaining the remodeling and restoration practices that follow.

No less nostalgic are claims by preservationist homeowners expressing a long held fondness for older houses, having lived in them when growing up or, also common, having lived in one during college. In fact, some of those interviewed said that it was while they were in college that they became conscious of their attraction because older housing was all they could afford. "And I lived in a bungalow court which I didn't even know that was what it was called at the time. But I thought it was charming, and I loved my little tiny bungalow" (Karen, Pasadena). Others are able to make the direct connection between the style of the house and personal memories.

> I moved out here from Pennsylvania to go to grad school at Cal Tech. And I was renting a house on Chester near Colorado and I really loved this house. I didn't know why I loved this house, because I didn't know anything about architecture, but it made me feel—it just reminded me of my grandmother. (Laura, Pasadena)

The individual's experience of living in an old house may be strongly affected by the social context of other family members' interests and sentiments. Several of our homeowners recall having lived in an old house at some time during their childhood, but admit that they did not appreciate its qualities at the time. Fred from Pasadena states: "I spent nine or ten years of my childhood living in an older house. But I wouldn't say that I got any appreciation from it … because I think my parents chose it for its price, not because it had character." Fred's father still does not understand why his son and his partner bought an old Craftsman bungalow:

> He [Fred's father] just bought his house directly from the developer, brand spanking new. Nobody had ever lived in it, and he thinks that's ideal. It comes with a warranty and everything is clean and fresh. He doesn't understand us at all.

Bart, Fred's partner, on the other hand, traces his appreciation of old properties to after his parent's divorce, when he moved with his mother to a rented 1906 house in the Santa Cruz Mountains of California. He comments that it was "a rustic farm home" with Victorian details: "We appreciated the fact that house was very old."

The experience of living in an older house as a child can play an important role in teaching future preservationist homeowners what to expect in terms of material conditions, and how to go about restoring the original qualities. When asked why she acquired her 1926 colonial-craftsman revival bungalow, Julie from Monrovia recounted growing up in an eleven-bedroom 1926 Dutch Revival house in Murfreesboro, Tennessee, in the 1950s. "We moved in when I was 7, so I was living in the house when all the repair work was going on. And it was in really poor condition." While Julie's parents did most of the remodeling work themselves, their goal was to make it livable, rather than to engage in "restoration"—an idea that garnered little civic attention or support in those days. Julie comments:

> People thought you were really strange that you didn't want to go buy a new brick ranch house. My parents knew exactly what they wanted—a nice piece of property and close to college. I walked to elementary school. I walked to high school. I could walk to church if I really wanted to. My grandmother lived 2 blocks away in another bungalow—a brick bungalow that she had lived in for 40 years … My

other grandmother had a bungalow, too. Most high school girls are cutting out pictures of wedding dresses and planning their future wedding, and I'm cutting out housing things from the paper with floor plans when I am 14. So, [mother] wasn't surprised when I bought an old house.

The fond childhood experiences Julie recalls now, however, clearly affected her adult interests and desire to acquire and properly "restore" an old house. While she knew she wanted an old house, as a single woman she had practical concerns, preferring something small to maximize affordability and living in a safe location near work. Julie's childhood experiences with her parents' can-do remodeling efforts contributed to her practical knowledge and confidence to take on the restoration project of her own home.

Of course, some homeowners also mention that they had grown up in a modern, or "mid-century" ranch or tract house, or had lived in both pre- and postwar houses. For those who grew up in a modern house, a family relationship sometimes stimulated curiosity about old houses, or triggered a yearning, a desire for an historic dwelling. One homeowner explained that her mother's acquisition of antique furnishings and periodic renovations of their modern tract house later inspired her desire to buy and restore an old bungalow. Similarly, a man, who grew up in a postwar ranch house, helped his father with weekend carpentry projects, and this stimulated his interest in woodworking, craftsmanship, and how things were put together. Today, he is involved in a long-term project of restoring a Craftsman bungalow. Another woman proclaimed that she felt she was "missing something" by having lived only in modern housing, and that old houses are "cool."

While the decision to buy an old house may be due to nostalgia for the past, many homeowners mention being captivated by the physical qualities of the old house. This attraction may occur on the first visit during the house-hunting phase, although homebuyers do not all seem necessarily attracted with the same intensity. Those who experience a strong attraction identify one special feature "grabs" them, and it is almost always an interior rather than exterior feature, as the following homeowner, Margaret from Alhambra, recounts:

> And so we started driving up and down the streets together and separately and we spotted this house on the market and we almost didn't come in—because it looks like nothing from the street. And it looked as if—I mean I think we've helped it along by painting it a more appropriate color. It used to be painted Easter-egg blue. And with battleship grey trim—it was weird—yeah, it was sad. And it was for sale. Do we really want to go inside? Well, what the heck. So it was springtime and the trees had leafed out in front and we came inside and we saw the china cabinet—and we said, "whoa, this is pretty cool" and I was especially ... well, it made me feel like I was in church. It reminded me of my childhood.

Fireplaces and still intact built-in china cabinets are frequently mentioned by homeowners as some of the first and most pronounced features that attract them to an older house.

Figure 4 Restored fireplace. Credit: Denise Lawrence-Zúñiga.

The wood craftsmanship, even if it has been painted over, and leaded glass or original hardware used in the dining room built-ins stand out as memorable features.

Question: So, what made you decide on this particular house?

Answer: I can tell you—it was that, that dining room built-in cabinet. Really, I just couldn't stop thinking about that dining room cabinet. (Laura, Pasadena)

Figure 5 Built-in china cabinet. Credit: Denise Lawrence-Zúñiga.

Figure 6 Old oak front door. Credit: Denise Lawrence-Zúñiga.

The recognition of craftsmanship in wood details, missing in contemporary houses, attracts attention and motivates contemplation. Luis, a Mexican–American homeowner from Alhambra married to a Taiwanese–American woman, bought a modest Craftsman bungalow in Alhambra in part because of the craftsmanship of the front door. Here, he describes with pride a family moment:

> And then my wife's grandfather is from Taiwan and he came to the house and he just sat, stood, and marveled at the front door. He just looked at that door and said, "That is a door! Do you realize what kind of material goes into that?" I mean he is 80 years old and his son bought a house in Palo Alto—a little cottage at the time for half a million dollars. And he was telling him, "You [paid] that much for your house, your house is a bunch of junk. This is a house!"

While most preservation homeowners tend to be middle class whites, Luis took some time to explain that his perspective had been influenced by a professor of architectural history at the local university, where he had previously worked as an administrator. Luis also recounted how his native Mexican uncle, who now lives in Texas and works in construction, proclaimed his amazement, when he saw the front door, at the challenge in hanging something so heavy. Luis proudly added: "People who know about this stuff or appreciate it, they stop at the door and [say] 'wow'."

Whether the contrast is explicit or tacit, homeowners recognize that older homes are "unique" in ways that mass-produced tract houses in new suburban developments simply cannot be. Luis commented that initially he and his wife considered moving to the city of Walnut. They had looked at five or six newer houses with a realtor. He says:

> They were so plain. They were so plain and similar that everything looked the same to us as we went from place to place. And I asked [the realtor] and he said: "Well, all these houses were built in the late '70s and '80s, and they all have these features, modern kinds of things." The doors were hollow, and the walls you could hear from one room to another, and here you can't.

Tom, from Monrovia, another owner who grew up in a GI tract house commented on them:

> I mean, they were mirror images of the one next-door and so flip, flop, flip, flop down the block. And so I liked the idea of having a house that not everybody had a mate to—an identical one. So, when somebody comes in here and says, "Oh, this reminds me of the house my grandmother used to live in," to me that's a compliment. It's a compliment because it's a different kind of house.

Many homebuyers stress the "ah-ha" moment of realization that a particular house is their destiny; their ownership of it seems inevitable. Some express this as an intangible as did Margaret from Alhambra: "And there was something too about it ... it was a kind of good energy. When we walked inside the energy was just right. I just felt good in there."

Homeowners fall in love with an old house because it is different, or charming, and it stirs an emotional response, or recalls cherished memories. Some homeowners imply that their house chose them. While first encounters tend to be emotional or even spiritual moments, living in an old house highlights the challenges of making it accommodate contemporary lifestyles while respecting its antique qualities.

To know the "original" house is to dig deep …

Everyday living in an old house quickly imparts a certain amount of knowledge to the occupants about its accommodating qualities, frustrating shortcomings and quirks, but it also presents mysteries that invite investigation. Most old houses require some renovation—the more deteriorated or neglected the house, the more work is obviously needed. Initially remodeling begins with cosmetic quick fixes like restringing stuck windows or covering walls with a coat of paint, a "white abatement" project as one homeowner called it. Deciding to completely gut and reconstruct the interior, however, requires deeper investigations of the house's material qualities. The assessments usually focus on house infrastructure such as identifying water, sewer, electrical, heating and structural problems, before stripping paint and wallpaper, or probing the house for post construction alterations. Making the house livable is the homeowner's primary motivation in generating an understanding of the house, but their initial discoveries just seem to provoke more curiosity about its history and material character, and provide avenues for deepening an emotional attachment to the house.

Homeowners repeatedly mention the "original" features of the house, and many seem preoccupied with discovering them, for conservation or restoration purposes. The concern with "original" qualities also dominates local civic discourse and professional approaches to the preservation of historic architecture, so these homeowners, who are all amateurs, inhabit worlds that are tangential with other influences. In fact, professional standards establish local benchmarks for restoration efforts and may also guide homeowners' individual efforts. For amateur and professional, the principal questions about origins include, when was the building built, who was the original designer and/or builder, and how was the building originally designed and built—that is, what are its distinct physical characteristics or style? Consideration of what material features are "original," and what features have been altered or replaced, provides the pivot point for homeowners' decisions to keep, throw away, conserve, rehabilitate, or reproduce architectural elements. Remodeling an old house does not necessarily imply restoring it to its former glory, but any investigation into original features, both archival research and hands-on, is likely to increase a homeowner's appreciation for the particular characteristics of a house.

Homeowners' research on historic identities is the main vehicle for establishing social relationships and potentially activating the material agency of the old house. The most critical information preservation homeowners initially seek is the name and identity of the designer or builder who built their house, and the original and subsequent owners and inhabitants. In some cities and historic districts, this information may already be well known and documented in local neighborhood historical accounts. Other sources of information may be found in archives and publications at the city's main library and the historical society, and in records of prior surveys, building permits and tract maps at city hall. Most homeowners have to do some digging for information, and there are many occasions where records are missing or incomplete. The identities of designers or builders are particularly important in the remodeling process because, if homeowners can locate other houses built by them in the neighborhood, they can get insights into how their own house was originally designed and built. Homeowners often ask their neighbors, or search old phone books housed in the library for more documentation, to fill in the gaps about the lives of previous occupants of their house. Some construct a genealogy of prior owners and occupants who passed through to flesh out the "biography" of the house. They may say that the biography belongs to the house, suggesting the house has its own autonomy and that the house endures while occupants successively pass through it. Much is made of the lineage of "ancestors" and their contributions in homeowner narratives. In some cases, homeowners make contact with the previous owners or their descendants, and may establish lasting social relationships with them.

Those homeowners who eventually move beyond remodeling to restoration or preservation frequently have the most complete knowledge about the designer and builder of their house and the date of construction, but they may also know what the house looked like when it was built. Fred and his partner, Bart, knew that their Craftsman house had been designed and built in 1912 by the Kieft and Hetherington firm, which had built a total of eight houses in their neighborhood. They were able to obtain a copy of a photograph from 1914 of their house, part of the "Flagg Collection" of fifty photographs of the neighborhood, now owned by the Pasadena Historical Society. The photograph, which hangs in their dining room, shows how the front of the house looked with the original landscaping, and that inspired the couple's ideas for restoring the front façade. They were also able to verify independently the date of construction and display the documentation next to the photo. According to Fred:

> The other thing that we found when we were doing some remodeling work is that Los Angeles Examiner newspaper from January 12, 1912, that had been inserted in the wall between the dining room china cabinet and the wall. It talks about Andrew Carnegie on the front page … We know it looks terrible, but we took a couple of pieces from it and had them mounted and preserved.

Another source of visual information for homeowners comes from other houses in the neighborhood that were designed and built by the same firm as their own. Paula from Pasadena describes how her plans to make changes on the exterior of her house had come from examining other houses designed by the same architect:

> We drove around last weekend and I have a checklist from our neighborhood historian of all the Irving Speicher houses in the neighborhood. And we drove by every single one of them. And we figured out that he had like 2 styles and we figured out which one was ours. We found another porch that, based on the scars of what looks like used to be river rock on our porch, we found another house with a river rock porch. So I think we know what we're going to go for.

This strategy is also useful for understanding how interior features were originally designed, especially where critical features may have been removed. After visiting several other Kieft and Hetherington houses in the neighborhood, Fred described his own dining room, and what it would have had if the previous owners had not remodeled it:

> This wainscoting, we are pretty sure, is original, except that that cap rail is not original. And it would have higher—quite a bit higher—and the plate rail is missing ... We are very certain that was the case because all of the other Hetheringtons have that feature. They have similar wainscoting and it goes up there to the plate rail. So we know that was what this would have had.

When there are enough local examples of a particular designer's houses to visit, homeowners can trace the evolution of style by reconstructing the sequence of construction and place their own house in the sequence. For instance, Laura from Pasadena and her husband knew that builder John K. Johnson had built their 1909 "Transitional Victorian-Craftsman" house. Laura says:

> He built about ... 12 other houses in [the neighborhood] and we have been in many of them and it is interesting to watch how his style evolved. This is one of the earlier ones that he built ... And some of them go into really truly Craftsman type houses, even though this one starts out as a transitional style house.

When there are no other houses in the neighborhood identified with the original designer or builder, or the designer cannot be identified, homeowners look for and study comparable houses based on style and age of original construction. Luis suspects that his neighbor's house which had minimal modifications also might have been built by "the same builder because there are the same features" as his house. Dan from Pasadena identified the spot in his living room where he believes columns had once created a divider with the dining room. "Mary, who lives on the next block in a house similar to this one, but two stories ... let me go in and take pictures of the ones in her house. And so, I was hoping that I would be able to start on those this summer—to reconstruct them, to put them back." Homeowners frequently consult with sympathetic neighbors to share information, but also attend home tours, visit architectural museums and exhibits, and read extensively to learn about the original style of architecture that their house exemplifies.

While many homeowners are able to name the designer or builder of their home, about half said they did not know or had not committed the name to memory, although they often had files in which they kept that information. The designer's identity is often thought to be critical to homeowners who aspire to restoring the original qualities of the house—a well-known designer confers some bragging rights and compels respect, but can also provide access to more information about design strategies and formal qualities typical of that designer. This knowledge is essential to homeowners who make inferences from formal qualities about the original design details, but also use the inferences to understand the designer's intentions.

The stories houses tell of designers and previous owners …

The main work of preservation homeowners is investigating and recovering the physical characteristics of their houses with the goal of restoring as many of them as possible to their original state. Knowing the identity and other works of the designer or builder is critically important to reconstructing the designer's intentions, but there is no substitute for the hands-on encounter with the materiality of the home to help discern those original conditions. Indeed, because of the passage of time and occupation by prior inhabitants, no preservation homeowner expects to find pristine conditions. Even if there has been no previous remodeling, there are likely to have been some changes just due to wear and tear and general deterioration. Preservation homeowners say their ideal house for restoration is exactly the one that has undergone little or no remodeling—a rare circumstance. Most prewar houses have endured interior and exterior alterations, which preservationists universally characterize by the severity of the loss of original features. Some prior remodels are devastating because key elements have been permanently removed or irrevocably altered, while other changes are considered superficial and can be easily corrected.

Descriptions of the physical conditions and features homeowners encounter and uncover make it clear that the material agency of the house is activated by the knowledge of the designer or builder, or prior occupants, the extent to which the latter are known. Generally speaking, the first owner occupants of the house are usually considered the most benign, and perhaps revered, especially if they also commissioned the house to be built or had lived there for a long time. Paula from Pasadena recounted that her modest 1922 house is pretty much as it was built: "Most of the details are intact … So, [the original owner] built the house and lived here until 1949 when he died, but his widow continued to live here until, I think, the '70s … She lived here for 50 years."

Homeowners presume that the original occupants would have little reason to alter the house and that the original conditions would match most closely their desires. The more numerous the successive occupants, however, the more probable remodeling efforts have cumulatively produced "inappropriate" results and permanently removed key features. The most ideal house is one where no remodeling or modernizing has occurred, about which homeowners brag. As Karen, a Pasadena homeowner commented, "We sort of thought benign neglect is in many ways better than somebody redoing, remodeling the wrong way—because you can always bring it back."

One way homeowners describe the remodeling efforts of previous occupants is to use an anonymous or indeterminate designation, "somebody" or "someone," to identify unknown actors. "Somebody had filled the area below that railing and had covered over the pickets up above and put in the big glass window" (Dan, Pasadena). Or, "somebody put oak floors over it so you can't see the scars from that removal of the floor heater vent …" (Karen, Pasadena). Attributing conditions to the anonymity of renters also appears in homeowner narratives: "So, it had been a rental for 30 years and it had been treated badly" (Julie, Monrovia).

Karen and Jim from Pasadena described their house:

> It was a rental for a long time. It was owned by a single woman who owned a couple of houses in the neighborhood, but it fell into disrepair and there was a list of code violations—they were almost going to condemn it.

While indeterminate designations are used to characterize positive and negative restoration efforts, most preservation homeowners are interested in a much deeper understanding of how their house was remodeled and why. One reason is that their experience with the house is three-dimensional—its forms and qualities impact everyday life—so that insights into the design intentions of original builders or later remodelers aid their construction of a meaningful interpretation—a consciousness—of living in a different era. Collecting biographical information about previous occupants and their experiences serves the construction of a meaningful biography of the house. Using anecdotes and stories gleaned from neighbors and information taken from library and municipal records, homeowners re-construct chronologies of prior owners to track and interpret the physical changes they made. As Laura from Pasadena admitted: "It really fascinates us how this house was lived in!"

Preservation homeowners' curiosity may lead them to make contact by phone or mail with former owners or family members. Tom and his wife were unable to identify the architect of their home. The awkward placement of the front door and relatively new construction materials surrounding the door created a puzzle. Tom from Monrovia was able to locate two previous owners and one was able to help:

> He was able to give us some insight about when some things were changed. For example, the front door wasn't originally where it is now. In a Craftsman home you would walk up the walkway, walk up the steps, and walk directly to the door. And so the door was moved to the side.

Laura from Pasadena and her husband learned unexpected things about their home from previous occupants.

> And we did make contact with one family that lived here in the '60s, and they had four kids and two of them have come by to talk to us. In fact, they told us that there had been a separate little house—a chicken coop—that had the same style of the house. It looked like a miniature of this house in the back yard, but it was lost. It was lost because [the son] was trying to cut the tree in the backyard and the tree went the wrong way and smashed the chicken coop. And then there was also some kind of turret thing in the front window that was also removed. So we get little bits and pieces from some of these people. But I think its time to get these people together again because we have never managed to get photographs, which is really the thing we would love to have.

Laura said she sent letters out to families, "Some of them have siblings who are still local, or one guy came to town for a family reunion and came to visit. You know, people just knock on the door to visit and say, 'I used to live here and want to see the house.'" Current and former residents sometimes feel a kind of kinship with one another created in large part by sharing in the lived experiences of the same home.

By talking to the neighbors and consulting some archival information, Margaret and Suzanne from Alhambra were able to develop detailed information on the three previous owners of their house, which was built in 1921. They constructed a fairly complete story of prior occupants' lives in the house, which informed their understanding of some of its features. Margaret says:

> The first owner was Samuel and Amelia Mays—it's really interesting to me because Samuel was like 55 years old when he first bought this house in 1921. So, they came here to retire. Amelia was an immigrant from—let's see she was Bohemian, so that's now part of Czechoslovakia. He was 55—he had sold farm equipment. He was born somewhere in the Midwest and they came here from Illinois. She was his second wife. And they didn't have any children … and when they first moved here they had other people living with them. It's a two-bedroom house, one bath, and they had the Harrisons and I can't think of their name … a husband and wife and his parents. So there were six people living in here until the Harrisons had a house built for them up the street someplace.

As Margaret explained, the Mays both died during World War II, and were not associated with any big changes in their house, but the Crowleys were the next family to buy the house and they built a mother-in-law house next door for the husband's mother. When the mother died during the 1980s, the Crowleys divorced and the husband moved into the mother-in-law's house. One of Margaret and Suzanne's neighbors told them that she knew their house as a child when the Crowleys owned it, and that the main entrance was then through the French doors in the dining room that faced the mother-in-law house. Margaret says:

> So the original owners were or at retirement age, and then the Crowleys raised their family here. I know he worked for the country doing measuring, weights and measures. She was a stay-at-home mom. We know that she was an alcoholic, at least in the latter part of her life. And then we know the next people were students and they probably didn't have much money … they got to it when it was time to sell. So she said that the dining room … was a dark green like a chalkboard. And, yeah, in fact, when we took the wallpaper down, that's what we found—dark green like a chalkboard.

Margaret described the task of stripping wood in the dining room as a revelation of how prior occupants had inscribed the house with their presence:

> So, we know from doing the wood stripping, that's almost a forensic project, because it comes off in layers. So the latex paint that they (the most recent owners) put on is gone in an instant. And then you get all the way down to the lead based primer, and you go over it and over it and you can see what each person did.

A similarly detailed account about paint comes from Jennie from Pasadena and her husband, Luke, who bought their 1910 Chalet-style Craftsman bungalow in 1985. Her first impression of the house was that it was too big for just the two of them, but her husband was a mechanic and really wanted the house with its five-car garage.

Figure 7 The celebrated multiple garages. Credit: Denise Lawrence-Zúñiga.

She describes the condition of the house when they first saw it and thought it needed "cosmetics."

> Everything was painted. The guy that lived here before was a painter. So we hear from the neighbors that he had tequila parties and would bring his buddies over with the extra paint that they had used. Everything—like this wall was light blue, this one was dark blue, the beams were all painted, the fireplace was all painted. Everything was painted.

Jennie spent years of painstaking work stripping all the paint from the wood, admitting that if some white paint remains, "You can cover it up with a little brown paint—no one will know." She had also heard from the neighbors that the painter's kids were troublemakers and that they had burned down part of the garage at one time. "And there were locks outside the bedroom doors where the parents had locked the kids inside their rooms. It was all beat up, but there wasn't anything that couldn't be straightened out."

Understanding the logic of earlier renovations is important for revealing the depth and breadth of alterations made on the original structure. One of the biggest challenges, which required a huge "commitment," was the de-stuccoing project on Fred and Bart's 1912 Craftsman bungalow, which they bought in 2002. The shingle house had been covered in stucco and was not considered a "contributor" to the historic district where it was located. The couple bought the house with the intention of "de-stuccoing" it after looking under the flared bottom of the exterior wall. Although they had not yet established the entire sequence of owners, they had identified the family responsible for the stucco job, in large part because the neighbors still remembered them. They were the Garza family who had moved into the house during the 1960s, but even before the Garzas, another family had modernized the exterior in the 1950s by nailing asbestos shingles over the original redwood "barn shake" shingles. Fortunately, before applying the stucco the Garzas removed the asbestos shingles, but did not remove the original shingles. Fred described the destructive techniques the previous owners used to prepare the exterior:

> And we knew from talking to neighbors with sister houses and from some under-lying evidence, that Kieft and Hetherington (the builders), their trademark was to put sort of a little bit of a Japanese flavor to it. So our wood trim is tapered, the shingles are tapered and we had—I don't know what you call that board, I always forget—oh, it's called a belt board. A belt course, so we could tell where the belt course had been … Just about everything below the belt course had to be replaced, but everything above the belt course, which is ⅔ of the wall, was mostly salvageable. It was clear that the belt course had been ruined to put up the asbestos shingles. Whoever put up the Sears siding, went along the entire belt course and took a wood chisel—because it stuck out about ½ inch—put it on top of that thing and bashed the front off so that all there was a splintery half of a board all the way around the house so that they could make it flush with the shingles.

When it came time for the Garzas to stucco the exterior, they tore off the asbestos shingles; Fred opined:

> I think at the point they decided that they wanted to stucco right up to the window frames. So, they went around and tore off all the window trim, and threw it out—all these beautiful, original wood pieces. And then they had to put in this stucco channel around that they just brought the stucco right up to the edge.

Fred argues that the stucco probably preserved the redwood shingles, but they were full of nail holes from both of the previous remodeling efforts.

Not only was the destruction to the exterior substantial, but as part of the remod-eling the Garzas undertook, they "ripped out almost all the wood windows and put in aluminum windows, and they put in really cheap ones that were really falling apart … And we have been slowly undoing all of that."

Although preservation homeowners bemoan the destruction or loss of defining characteristics of their houses, they do try to understand the reasoning behind it. Fred and Bart explained that the modernization of their house was characteristic of the historic period of the 1950s itself. Fred describes the remodel of their fireplace:

> And all this sort of fits with the 1950s view of the craftsman that they were the worst … They were the least popular style at that point, … and the colonial style was really popular. So, people were going through and stripping a lot of them out. So this is all sort of a colonial redo.

Figure 8 De-stuccoing around the doggie door. Credit: Denise Lawrence-Zúñiga.

Figure 9 Shingles restored on the de-stuccoed house. Credit Denise Lawrence-Zúñiga.

Even when the previous owners have engaged in some questionable design decisions, preservationist homeowners say they appreciate any prior efforts to restore qualities of the house. Specifically, homeowners identified essential repairs to the plumbing, electrical upgrades, heating, and air conditioning systems for which they were grateful. As the sixth owners of their house, Karen and Jim from Pasadena know that the immediate previous owner, Paul, and also a university professor before him, had tried to restore or rescue some features. Jim said, "Paul seemed to really like the house and to want to make it nice, but he just didn't have the money." Karen added:

> Well, … we could see Paul through the windows repairing the fireplace because we actually rented a house down the street—that's how we ended up here. But he had put a kind of a drywall box around the fireplace and … we bought as is. So we knew that it was probably really screwed up.

But then she credited the professor with stripping the paint from the wainscoting in one room. "I mean, luckily, somebody, [the professor], wanted to bring it back at some point."

Homebuyers who are likely to become engaged in restoration work usually look for some original qualities when considering their purchase. Lifting up carpets to see if the original wood floors are in good enough condition to be refinished, or checking to see that original interior doors or cabinetry remain, are minimum steps to ensure that the house is in part salvageable. Once homeowners move in, the most frequently mentioned missing items are door hardware and original light fixtures. When asked about it, Karen replied, "Inside the house it was turned into cheap stuff—glass knobs—that you find at O.S.H … But the window hardware, that's all original. And all the windows are original to the house." Particularly aggravating is the last minute "upgrades" made by the previous homeowners just to sell the house. Margaret from Alhambra complained that when they first moved in:

> Yes, it was pretty livable; it was just ugly. Before we moved in, we actually did some work for a few weeks on it—painting—stripping off wallpaper. Sadly, we took off wallpaper that they had put up to sell it … It was ugly. We took that down. We were dismayed that they had this tile that looked like it came from Orchard Supply, which her brother had badly tiled. To add insult to injury, they put in the wrong tile and laid it badly. And they had beautiful little octagonal tiles on the kitchen counter that they took out and dumped in the back yard, which I found—I was dismayed and we were sad.

Very often houses have fallen victim to "unfortunate" modernizing efforts, one of the most common is the replacement of original double-hung or casement windows with aluminum or jalousie windows. Fred recounted his discovery of the original windows in the sleeping porch which had been filled in with inexpensive aluminum jalousie windows because, he suspects, the owners did not know how to repair the old windows.

> And someone clearly didn't know at all what they had or how they worked, because, I guess, a long time before the ropes had rotted and the weights were just laying in the bottom of the wall … Because the way ours work is that we have—I don't know what you call it—it's this piece of wood that slips up and gives you access to slide the window down and when it slips down it gives you a window sash. But, on some of the windows, that was down and the window was closed on top of the wood. It's supposed to be up. On others it was pulled up and they just painted it … So we are looking at it and saying what is this funny little wood piece and why is it on this on this window and why is there an aluminum window here? We couldn't quite figure it out.

Other modernizations retain some features, remove others, and add inexpensive modern materials to the mix creating incongruous stylistic combinations. Karen and Jim described the bathroom they targeted first for remodeling—it had its original medicine chest minus its door and a claw foot tub, but the original sconces had been replaced with a big 1970s mirror with "show lights" and inexpensive "modern" faux tiles that were now deteriorating. It was difficult for them to "salvage" much of anything. Despite the loss of many original features due to prior remodeling efforts, homeowners are overjoyed to find the original doors, windows, screens, light fixtures, or hardware safely stored, and often forgotten, in the basement, attic or garage. In fact, these "finds" become legendary in the preservation community and are included in narratives about old houses because they imply previous owners didn't know the value of what they had. Bill from Alhambra says his windows are "cool" and his friends had never seen any like them.

They were attached to the screen and they go like this ... Some were rotted out, and they had put in slotted screens. Luckily, when they put in the deck, the windows must have had all the hardware and they stored the original screens and hardware in the basement.

Some houses, however, are simply a disaster—they lack heat, the plumbing is broken, the plaster is crumbling—they may be actually dangerous to live in even though the dedicated preservationist does so. Jim and Karen had bought such a "fixer-upper" and within a year they made a dangerous discovery. Karen comments:

And then one night I felt hot water coming out of the wall—through the wainscoting ... I was just walking in the dark and ... using my hand to find my way along, and there it was hot and wet. So I pulled off the wainscoting—the batten—and there was a pipe there. And it turned out that our water heater vented up into our old kitchen chimneystack, but there had been a gas stove there. But they never plugged it back up, so the steam was coming through the wall every time we turned on the hot water. And the carbon monoxide was coming in. When the plumbers came to look at it, they couldn't believe it—they said, "you could be dead."

Finding clues to the original design often entails many days, entire weekends, and even vacations of painstaking work. Homeowners often talk about the "scars" or "ghosts" of original design features that, despite patching efforts, can still be identified in the floors and walls of the house. When describing his and his wife's efforts to restore the fireplace, Jim said, "We just kinda guessed—we could see outlines on the ceiling and on the wall, so we could tell the basic shape. And we kinda guessed what it might be. And we worked with input from our friends, and that's what we came up with."

Similarly, Julie from Monrovia's experience with removing a large 1950s plate glass window revealed evidence of the original windows. Her carpenter noticed the number and their position, saying:

Do you want to hear something funny? They used the same frame when it was a smaller house when they reframed the window they took the frame out and just replaced the windows. The ghost marks are here—it's exactly the same size, even where I am putting the hinges and latches; it's identical to what was here before.

Jennie from Pasadena pointed out the "ghosts" of the crown molding long ago removed in her living room and entry: "We just use this room as a den—you can see the ghost patterns for the molding in here (entry) and I had my brother-in-law make new ones in here (living room)—the crown moldings. So he put them up for me."

Much of the discovery about the identities of prior owners and occupants and their impact on the house contributes directly to homeowners' construction of narratives inclusive of relations with these personalities. This knowledge is added to the documented information about the original designer or builder of the house. Although most former occupants have passed away, and certainly those still living have moved away, the homeowners construct a kind of genealogy for their house that recounts the chronology of designer and resident personalities and their biographical

Figure 10 "Ghost" from prior construction. Credit: Denise Lawrence-Zúñiga.

information linked with specific material features and alterations to the house. If the identities are unknown, homeowners still try to piece together a plausible scenario to explain the physical features of their houses. Ultimately, the goal is to try to uncover the original features, to discern what they were from altered remnants and ghosts, which inevitably entails contemplating the intentions of the original designers but also the previous occupants who remodeled the house.

What family-connected homeowners know …

For homeowners who grew up in the old house they now own, or who own the home of one of their own family members, the experience of the "original" house is grounded in much more personal memories of family and place. Two homeowners who grew up in their houses live in a 1887 Queen Anne and a 1909 Craftsman, respectively, in Ontario, while the one who was able to buy his grandmother's 1887 Eastlake house is in Monrovia. All of them have strong feelings about their houses but also talk with great pride about their family's role in the founding of the city, local politics, and/or local business. Living in and restoring the house is a way of honoring their memory. Moreover, their houses clearly act as vehicles for specifically recalling cherished memories of family life experiences. Stan says initially the 1887 Monrovia Victorian house, which he is restoring, attracted him because it had an "overwhelming sense of family" for him. He is able to trace his family to the founding of the city and his great grandfather who had moved from Michigan to Monrovia, in 1888, and bought the then newly built house. He also knows the house was moved to its current location in 1900 after his great grandfather died and he can trace many of the alterations made to the house by his grandmother who lived there until her death in 1966. He described how she had subdivided the house into apartments in 1939 to earn a living, and how he had spent time with her when he was growing up and had a chance to learn about how the house was inhabited.

Stan is recognized as a local city historian, a role he embarked on when he moved back to Monrovia in 1967. Stan's passion for history led him to expand his interests beyond his own family's house to research the city's founding families and its earliest houses. In this context he confirms the attribution of agency to the materiality of the house noted by other homeowners saying:

> I have this notion, that I cannot support with anything, that people who live in houses impart part of themselves to the physical structures. They impart psychic energy, or something. I think they leave a record of the fact that they have lived there. So, I think it is important to tell their stories.

Sarah also conveys that deep sense of family when talking about her 1909 Craftsman house in Ontario. The house has been in her family since 1918 when her grand-parents, who owned citrus packing sheds, bought it and the surrounding orchards. She explains that her uncle had been mayor of Ontario at one point, and that her father, who never lived in any other house, loved the community. Eventually he bought up just about every house on the block. Although her family maintained another home in Newport Beach, they considered the Ontario house as their primary home because it was close to the family business. Sarah characterizes the house as a setting in which her father told stories, and the stories are what inspire her to preserve the house even though her father, she says, was lousy at maintenance.

> I lived in this house and was at my father's knees sort of—because he told stories all the time. At that time, being 45 [years old] and having a child was a big deal—not now. Because he was so much older and because he had lived his whole life here, he'd seen all the changes. He knew that where my elementary school was there used to be orange groves.

When her father died in 1997, Sarah and her second husband, Ben, a designer-contractor, moved from a 1926 Spanish Revival house they were restoring into the 1909 Craftsman. Sarah had previously worked with her father to rehabilitate and restore some of the older properties he owned, and she had furthered her appreciation of older architecture by getting a master's degree in historic preservation in the public history program at a local university.

> I've always loved bungalows, but then I kind of—I went to Scripps College—so then I loved Spanish style. So, really, I just love the lore of Southern California and those two styles are mixed up in it. There is just no way around it. I think I always knew I would end up living here and that's why—it was sort of an understanding between my dad and me that I would live here. So, I think I wanted to live in a Spanish style house because I knew I would wind up living in a bungalow. My destiny is here.

Moving into the old family house, however, was a bit overwhelming because, according to Sarah, her family never threw out a thing. "When we were moving in here, our biggest battle was getting rid of generations worth of stuff"; that included closets and bedrooms packed to the gills that hadn't been opened in decades. Although much of the "stuff" turned out to be junk, some of it could be given to the local history museum. The three-car garage, which Sarah had never entered the entire time she was growing up in the house, was filled with antiques, and a nearby packing shed was full of early twentieth-century Stickley and other kinds of furniture. As Ben observed about their first months in the house: "It was like everyday was a mystery!" Flush with family resources that had been safely and securely stored, Sarah and her husband were able to furnish their home with heirloom antiques.

George's grandparents acquired the 1887 Queen Anne in the 1940s, long after it had been remodeled in the 1920s, by a previous owner, who had added a mansard roof. His grandmother had moved to Ontario from Long Beach to set up a very successful candy store in the center of town, but her husband died shortly after the move. George's father and mother moved into the house to help his widowed mother run the family candy store business, and George and his brother grew up in the house. George says he likes history and older architectural styles—the Queen Anne is particularly appealing to him—and he sees the house as having roots in the history of the city. He recalls growing up in the house—his grandmother's bridge parties in the front room and her friends playing the piano. George also has fond memories of his friends coming over to the house, which they treated like a big "playhouse." He has amassed a number of photographs of the exterior of the house showing how it looked before the mansard roof was put on and he talks about bringing back the original design—about fixing it. As in Sarah's house, very little had been altered in the interior of the house—a popcorn ceiling, wallpaper and some painted woodwork—and most of the original details were intact.

These three homeowners had acquired more detailed knowledge by having lived in the house and from photographs, but also by knowing intimately family members who have since died, than other homeowners who were interviewed. While there may be gaps in their knowledge, they express a sense of privileged entitlement, of "ownership" that other homeowners lack. This is a kind of authentic and unquestionable connection to the house that almost no one can really challenge. I say "almost" because Sarah mentioned that her sister, who has never had an interest in the family home, got upset during a holiday dinner one year proclaiming, "This isn't my house anymore, this isn't the house I grew up in!" Sarah says at first she was upset, but then realized that it is no longer her sister's house because it is now hers.

Conclusion

Old houses, because they have endured so long, hold a fascination and invite attention and care, perhaps calling for their promotion and local recognition of their social and historic value. Their quaint features—the drop-down ironing board in the kitchen, the claw foot tub, the service porch with wash tub, the sleeping porch, the gravity heater, the milk delivery slot or the California cooler—all capture the imagination about lifestyles no longer directly accessible. The contrast between new homeowners of older houses and those who have inherited or purchased anew the houses they grew up in, however, points to the challenges presented to the former to compensate for a dearth of social and material knowledge about their house. The material qualities of home and their agency allow people to remember the life, or to recreate an imagined experience of life, in a different era, while enjoying its aesthetic aspects. The old house allows people to make new what was old and make it relevant and meaningful to a new generation of homeowners. Some of the agency of the old house is captured in homeowners' narratives by citing the names of recognized designers and builders, or the historical families who once lived there. Some of the agency is found in the specific design features and craftsmanship of a bygone era, or in the exploration and revelation of those features covered by years of remodeling. And some of its agency is found in the everyday life lived in a three dimensional space that affords and shapes the habitus of household members. Exploring the old house and uncovering its mysteries sharpens homeowners' discerning judgment about its original material qualities, while simultaneously deepening their attachment. The activation of material agency is only fully realized in homeowners' restoration practices through which the intended original forms are reinstated through the agency of the homeowner.

Restoration Strategies, Imagining the Past, and Reconstructing Historic Meaning

While homeowner efforts to discover and reveal the original features of their homes mark a serious commitment to the preservation enterprise, that work constitutes at most only half of what is needed to "restore" a house to its original glory. These homeowners say that proper preparation—the careful stripping of paint or wallpaper, de-stuccoing, making windows, doors and hardware work again—is the most important step in reaching restoration goals. Indeed, unmasking hidden treasures so that they can be appreciated and revalued may be deeply satisfying, but it is an incomplete story. Restoration strategies anticipate the production of particular material qualities that put back the "original" form and character of the house, as best as can be determined. Homeowners do not take on these tasks alone, but collaborate to produce the desired results. Remodeling decisions are conditioned by the social context of the household, neighborhood and preservation community, and by seeking direct help from designers and contractors experienced in preservation. Still, homeowners must know what they want to achieve in the remodel, and the preservation homeowner takes this task very seriously.

As homeowners become familiar with the material and social qualities of their own houses, they also discover the challenge of fitting their needs for living spaces, and especially contemporary lifestyles, into the older house. Reconciling occupants' needs with the material constraints of the house often requires adjusting behaviors and expectations as much as it requires changing house form (Miller 2010: 96). In many ways old houses become teachers to willing homeowners who learn their idiosyncrasies and gradually transform its physical properties. The relationship between people and their physical environment is dynamic and mutually reinforcing, and co-produces both material realities and social-cultural patterns (Bourdieu 1977). The physical qualities of the old house are a manifestation of generative cultural schemes and dispositions of the habitus that were simultaneously embodied in everyday practices during a previous era. The old house instructs the new occupant on how to live there, sometimes revealing in its curious forms and features a lifestyle set in the past, and provoking a negotiation favoring a new adaptation for living in the present. The material qualities of the house guide and shape daily routines, enabling and constraining individual and social behaviors, to change the occupants' lifestyles. The house is embodied space; as one homeowner metaphorically characterized her relationship with her house, "It wraps its arms around you."

The embodiment of social and material knowledge in the house becomes routinized as habit, it is "naturalized" and taken for granted, but it does not preclude the introduction of physical changes through individual agency. Homeowners in general are all likely to alter some features of their houses through remodeling, decorating, or simply performing maintenance on the structure. In doing so people compare their actual home to images of what they aspire to have, not so much to keep up with the neighbors, but to represent themselves to themselves as corroboration and reinforcement (Clarke 2001). Thus, the practices of renovating, restoring and preserving an older home construct identities in realizing aesthetic and functional schemes, but also invest the homeowner's agency in activating the agency of prior builders and occupants who initially produced the dwelling's material qualities.

The primary motivation for self identified "preservationist homeowners" in remodeling is the desire to return the house to its "original" conditions. This goal is also socially constructed, codified, reinforced, and reproduced as part of a professional cosmology of historic preservation to be discussed in the following chapter. As a popular remodeling practice, however, the homeowner's search for original qualities finds support among like-minded neighbors and community members, as well as from professionals, which serves to legitimize and produce value for restoration outcomes. Homeowners' practices include DIY-"hands-on" efforts, or hiring designers and contractors, or, most often, some combination of both. At the very least, the preservation homeowner makes the decisions about what original qualities are to be restored, and seeks knowledge about how to do it.

As we have already seen, discoveries and revelations bring homeowners into a social relationship with the house through its prior agents—anyone who had previously designed, built or altered the physical form of the house. Homeowners' narratives reveal the complex relationships, and often ambivalent, feelings they conjure up with dead or missing actors, and how the sense they make of previous occupants shapes their own identities and lifestyles. The actual practices involved in restoration— stripping paint and wallpaper, refinishing floors and surfaces, repairing windows and doors, tearing out walls and cabinetry, and the like—constitute the primary means for integrating the agency of others into homeowners' identities. Each active remodeling practice engages homeowners in bringing back to life selective features of the house and, in the process, constructs a restoration ideal and its rationale. Idealized uses are fashioned around the house's history to justify restoration strategies, but just as often they involve fitting selective historical facts to household members' needs and wants. The reliance on history to tell the story of the house and situate new homeowners within that history engenders in them a sense of stewardship for the material qualities that they have salvaged and rehabilitated. Stewardship expresses a kind of reciprocal relationship of protection toward the material world that anticipates future responsibilities assumed by generations of house occupants still to come. Of course, there is no guarantee that future occupants will hold the same sentiments as the restoration homeowners, but that is the hope. In these ways, home restoration practices—the acts of remodeling a house—are transformative, and produce identities and lifestyles with deep meanings for those who engage in them.

The restoration "ideal" is the central philosophical issue for preservation homeowners. Although some consider themselves or others to be "purists," there are significant debates about what this ideal should be. Homeowner narratives often reference the discourse about preservation ideals in order to position themselves in a social field that includes other household members, neighbors, and members of the larger community. The ideal centers on the degree to which homeowners express a moral obligation to retain, restore or reproduce the "original" features. They put in considerable effort to work out their own logic of remodeling in confronting this question. Some of the critical issues include choices between modern conveniences and the original amenities, especially in kitchens, bathrooms, and heating and cooling strategies. Other concerns are the extent to which original building materials should be conserved and reused, or reproduced exactly as craftsmen produced them originally, or whether spatial configurations that limit contemporary lifestyles should be redesigned or salvaged, which may require homeowners to adjust their own behavior. In many ways, the discussion of original conditions is a debate about how one "should" live in an old house—how much influence the material conditions of a previous era should impact everyday lives.

Homeowners' restoration ideals and interpretations vary widely and include recreating stylistically consistent or "period-appropriate" spaces, furnishings and reconditioned appliances, or mixing some original features or reproductions with modern appliances and conveniences, or emphasizing a comfortable and livable home while rejecting a "museum" staged with artifacts. For the inexperienced hands-on homeowner there is often a steep learning curve required to make an old house livable while achieving "preservationist ideals." Despite their best intentions, homeowners are likely to make "mistakes" through trial and error as they learn what is "appropriate." Theirs is a long-term commitment to restoration—it cannot be completed within a year or two; it takes time, focused effort, the acquisition of specialized knowledge—and, of course, money. Because there are likely many "unknowns" buried in the old house, specific restoration goals and strategies cannot be completely defined in advance, but develop and evolve as homeowner's remodeling proceeds and knowledge increases. Even among experienced homeowners the remodeling process takes time because each house has its own idiosyncrasies that require discovery and assessment. Negotiations over occupants' dispositions and the materiality of home may involve extensive experimentation and mutual adjustment.

Actively restoring a home's original properties requires attention to myriad minor decisions over years of sustained effort, but the production of a restored house also co-produces and integrates the homeowners' identity and lifestyle into a concept of "history." Here, the "original" features of the old house stand in for history; the artifacts carry both the associational meanings with people—designers, builders and previous owners and occupants—and events, and the ontological meanings that reference material qualities such as craftsmanship, durability and aesthetic-symbolic attractiveness. In remodeling and restoring the old house, homeowners construct their history experientially so that the house not only embodies the social relationships of their own family and friends, but also connects them to actors whose agency they have reactivated and restored in material form. Homeowners develop a protective stance towards the restored home articulated and justified through narratives that reconstruct the past. In these ways, home restoration practices are transformative and produce the identities, lifestyles, and deep personal meanings for those who engage in them (Lawrence-Zúñiga 2010).

Restoration practices ... experimentation and learning

Hands-on remodeling practices, whether they begin with stripping paint or de-stuccoing the exterior, necessitates a selection of material features to rehabilitate and decision about what manner. Some treatments may be technically determined such as identifying the proper type of paint, stain or polyurethane needed to preserve or restore the qualities of aging wood, but even these choices may be challenging because contemporary products do not use the same chemical ingredients as the old ones. For example, lead paint, is prohibited by current regulations. Moreover, many new construction techniques and regulations are incompatible with the material qualities homeowners wish to retain or enhance in their remodeling efforts. Most often, however, the choices that confront homeowners are aesthetic or stylistic first, and of a technical nature second. They soon realize they must acquire the knowledge necessary to make the "appropriate" decisions and seek out informational documents, educational workshops or videos, hire craftsmen and specialists, and seek advice from like-minded friends and neighbors.

A few first-time homeowners openly admitted that they had experimented with different—somewhat botched—attempts at remodeling when they first began. They describe their mistakes as part of a process of learning about both style and technique. Jim from Pasadena admitted that he and his wife, Karen, knew little about their Craftsman bungalow before they started renovations:

> I mean we knew it was an old house, but we didn't know what style it was, we had never been to the Gamble house,[1] we hadn't done anything—we hadn't bought a single book ... But I had no idea—I wanted to create the back room like blue Scandinavian style—something out of *Better Homes and Gardens* [I saw] when I was getting my haircut. But that didn't last very long ... I think we went on the home tour in the neighborhood in April, in the next month, and that's what got us started.

Jim and Karen explained how quickly they were able to learn the "correct" stylistic and technical approach by visiting restored houses in their neighborhood and, even though their budget was limited, they took immediate action. Jim went on:

> So when we went on the home tour we saw everyone had fir floors and we thought maybe have fir floors—they had been painted brown with brown porch paint, and then we painted another layer of white paint on them—when I was doing the Scandinavian thing—and it was coming off on our feet. And so, we saw that

everyone had fir floors, so then we saw the fir floors and that's what we have … we didn't have much money and they were the cheapest guys. So they rolled up in a van, like in a Cheech and Chong movie, and they spoke no English. They refinished our floors for like $500 for three rooms, and they are ok.

Other homeowners initially seemed clear about removing the obviously discordant items, but then lost their way, experimenting with different approaches that hadn't yet panned out. Bill from Alhambra explained that he got rid of the "tacky stuff" first in his Mediterranean Revival house by removing the saloon doors that had been added to an arched doorway. Then he started experimenting with paint and decorating:

> It's a problem because it's a Spanish house, I started to go crazy because I'd go to Mexico and I bought that carved statue and a few other things. I mean because it's the Spanish style of the house, I got carried away with too many angels, the sun and the moon. And then I realized that this is a 1930's house, too, and it probably wasn't a big Mexican house. I changed my mind a lot.

If a house is very deteriorated, the first-time homeowner may feel pressed to do something, such as painting all the walls to make it livable, before learning what preservationist approach to take. Learning to "see" the potential for restoration in the most derelict of material features is a critical component of the practice of preservation. Laura from Pasadena described her awakening like this:

> I'm glad we did all this stuff on the inside that you couldn't see [electrical and plumbing] first … because my ideas of what to do with this house were so wrong. I had this vision, because there was this house down the street that was white with yellow trim, and that was like my goal. Oh, I'm going to paint the house white with yellow trim … and I just didn't understand the whole Craftsman thing. And then we started learning, and started reading, but we really hadn't understood what it meant. And so by the time we were ready to do things that people were going to see, we then had a little more education and sophistication and understanding of what it is. That was probably a good strategy in retrospect. In retrospect, there were so many things where I thought I really wanted to do this and then realized that that would be the absolute wrong thing to do.

Considerations of style and period of construction loom large as novice homeowners begin to understand the house they inhabit. Constructing an idealized aesthetic framework to capture aspirations is central to organizing and guiding preservation practices. It shapes homeowner decisions about what spaces and formal features should be restored and how, and gives definition to the final product when the work is complete. Almost every homeowner interviewed could name the style and identify the period of construction of their house, and many could talk extensively about its various original stylistic elements or later period defining alterations. Descriptions of architectural styles such as Victorian, Craftsman, or Mediterranean Revival are codified in the architectural and preservation literature and many homeowners become familiar with the formal characteristics that differentiate their house from others. Restoration practices, then, often follow stylistic typologies when considering the particular parts of the house that need to be preserved, replaced or reproduced. Style categories also influence the specific period-appropriate furnishings or appliances to be acquired, not as isolated items, but as they fit together to complete an idealized aesthetic unity.

Even elderly homeowners, who have lived in their houses for decades without ever doing much remodeling, also report having misjudged the significance of some "original" features of their houses they became aware of later. Several reported having mistakenly thrown out glass doorknobs or cabinet pulls decades earlier because they did not know what they were, or did not think they were valuable, and they bought modern replacements they now regret. As they have become aware of their value, some have recently begun to search antique hardware stores and flea markets to find period replacements to restore the features in their houses. Other homeowners report making the mistake of tossing out pieces of wood stored in the basement that they later discover were the shelves to their kitchen cabinets or linen closets. Correcting these errors in judgment requires homeowners to perceive their houses differently, to pause and ask important questions about the function or origins of items before dismissing or relegating them to the trash.

Some first-time homeowners seemed to know the importance of taking time to become familiar with the house before they began renovating. Julie from Monrovia, who had grown up in a family that did remodeling, described her renovation practice as selecting design ideas based on the style of her house and matching them to the physical features she had uncovered. She used these to convey her desires to the contractor she eventually hired:

> I didn't do anything right away—I waited to live in the house and try to educate myself on what other homes of the period would look like by going on home tours, getting magazines, checking books out of the library. And I started a loose-leaf binder of things I thought I liked, or things I might want to do. Because I didn't have any money to actually do any repairs once I bought the house, I did the basics to live in the house—but I didn't have any extras. So, once I sorted out the kinds of things I liked ... I would look at pictures and think, well, there are ghost marks on the wall, there were bookcases, how lovely it would be to have a bookcase again. The ghost marks you could see in the plaster. Sometimes you ... get ideas from looking at other people's homes, what they've done. Sometimes its something you just think would look nice.

A few homeowners were already experienced in hands-on restoration practices because they had previously owned and renovated older houses before they purchased the one in which they currently reside. Before moving to Monrovia in 1986, Tom had owned a 1911 house in Alhambra where, for lack of sufficient funds to hire others, he learned restoration techniques. He says:

> I already knew how to do the plumbing from the wall in. I didn't know anything about electricity, my neighbor taught me electricity. So, I was able to start doing wiring and things. I didn't have woodshop in high school, so I took woodshop at Alhambra high school at night for a number of years, because they had the

tools—the major tools—so I was able to use those to do some project and learn how to use the tools, which I did. Then, I built all the cabinets in my kitchen there—I redid that kitchen. So, over the years I've been able to buy tools, I've used the PBS TV program—first *This Old House*, and then they added the *New Yankee Workshop* and *Ask This Old House* ... So, I've learned how to do things. So it's like a surrogate parent, a surrogate father, showing you how to use the tools.

Most homeowners, however, hire others to do electrical, plumbing and any major construction. This has stimulated growth in specialty businesses, contractors and craftsmen who cater to restoration homeowners and teach them technical aspects about old home repair.

Restoration practices ... balancing ideals with costs

Once homeowners have ascertained the number, types, and location of damaged or missing windows, doors, bookcases, breakfronts, cabinets, hardware, lights, and the like, they are confronted with decisions about the resources and knowledge needed to repair, replace, and restore them. Their strategies incorporate a variety of preservation philosophies they use to explain, and sometimes justify, what and how they restore and why. Homeowners' practices must balance their preservation ideal with their financial resources to pay for materials and/or services, and their own knowledge, time available and skills to do the work themselves. Some homeowners for reasons of economy argue that they have more control and can get higher quality workmanship if they build it themselves. For those with adequate financial resources but little skill or time, hiring craftsmen or a contractor is often the answer, but this does not necessarily guarantee high quality results. Margaret and Suzanne from Alhambra hired painters to strip and paint the exterior of their house, but weren't entirely happy with the results. Margaret comments:

> We wished they had done a better job. They were here for a month—we were living with them. It was just one guy who was half blind and his nephew. He can't see anymore. It's the paint dust and the sanding and it's kind of scary because it's his livelihood. He was really nice and he painted other things in the neighborhood, and he was unbelievably cheap. So we hired him ... And I wish we had money to hire someone who might have done a little better job, but it looks better than it did.

The trade-offs between expectations for quality results and resources to pay for that quality are central to the homeowners' strategies. Some do not have very high expectations for quality work and others, like Jim and Karen, and are willing to accept lower standards for selected projects because of a limited budget. For those with few resources but high standards, restoration can be a long-term effort spanning many years. On the other hand, there are some homeowners who are more concerned with creating a certain appearance or completing the project quickly, regardless of how it is achieved, than with the qualities of materials and construction processes used.

The differences in remodeling practices, expected outcomes and rationale in preserving the original house define the preservation homeowner's philosophy. Those who demand retention or a faithful reproduction of the original features are often labeled "purists," but there are many variations of preservationist "purity" to be found in renovation practices. One "purist" goal is to restore the exact material qualities of the house exactly as they were when it was first built. This would include retaining and restoring all its original features, as much as is possible, or acquiring exact antique replicas and using the same construction materials and methods. Embedded in this approach is a moral and aesthetic conviction about the importance of maintaining the material integrity of the house. Purist philosophy emphasizes the house should be kept whole and not corrupted by irresponsible additions or alterations. These sentiments are expressed in phrases such as, "the house was meant to have ..." or "keep the house true to its original design." Dan from Pasadena described his 1913 Craftsman as having this kind of stylistic integrity:

> Well, houses of this style are just really comfortable. That is, if they haven't been hacked up, if they are true to the original design. People think this idea of indoor-outdoor living arrived ... in the '50s, but it started in the teens, because all these houses have big porches and have other doorways, and they were not even really meant to originally have curtains over the windows. And, so the way the house is laid out, it just really works well ... I kind of feel like every room is just the right size and just the right location.

For purists, reproducing missing or unserviceable features by making them anew is considered a legitimate substitute for lost originals. When the original windows or doors are not salvageable, custom-made replacements are made to match by craftsmen who specialize in reproductions. Homeowners who produce the same quality for less money by making the items themselves, eventually becoming skilled craftsmen, as Jim from Pasadena did. In fact, carpentry and construction can easily transform homeowner identity and lifestyle when they become regular practices.

Before they completed their restoration, Jim and Karen added a carpentry shop: "We built a 1½ garage which is a little bit bigger than the original one ... And that's why we needed it because all the equipment was in our living room before."

The demands for faithfulness in recreating the exact original conditions may also lead dedicated preservationists to insist on "original" materials, rather than substitute contemporary products. Reproduction window frames, or replacement windowpanes, for example, should use antique glass harvested from houses slated for demolition and sold by salvage companies; the glass looks wavy and imperfect by contemporary standards. The use of original antique hardware if available also seems obligatory. Doors are a challenge because, if they are to be painted, new ones can be custom made, although people say they don't exactly match the originals. A preservationist preference is to find interior and exterior doors at salvage yards and stores specializing in antique building components; and sometimes people find these items online at eBay or Craigslist. Antique hardware and lighting fixtures are also fairly easy to find online and at flea markets, salvage and antique shops, and these are preferred to the "authentic reproductions" widely available at trendy shops, through

Figure 11 Kitchen remodeled with owner-crafted shiplap applied to the new refrigerator to the right. Credit: Denise Lawrence-Zúñiga.

catalogues and even at big box retailers. Cabinets, cases, and breakfronts typically must be custom made if they are completely missing and the materials for these should be the same quality and type even if new wood is used rather than salvage or original wood.

The concern with original materials can also take on an intensely moral quality among purist homeowners. After trying unsuccessfully to fix the only bathroom in their house with a coat of paint, Jim discovered that the sink was emptying under the house. He and Karen decided that the bathroom had to be their first project and they consulted with a number of contractors, but in the end decided to oversee the work themselves. Jim says:

> We took it all apart and all the wood and plaster. We knew a guy who is an old fashioned plasterer and we didn't dry wall—we got him to plaster it. And then, [Karen] stripped almost all the wood that we could save, then we bought—got some fir—we didn't want if somebody strips it someday for someone to be sad that there is no fir under that. We thought someone is going to strip it someday and find that piece of plywood, they are going to be all optimistic about it and then they're going to strip part of it and find fir, so they'll keep going. We're pretty much doing that on all the work, replacing the actual wood, clear fir. We don't want anyone to be sad. It would be sad.

Here is a clear illustration of the tension between choosing materials that match as closely as is possible the original construction materials, plaster, and Douglas Fir, and choosing a cheaper substitute such as drywall, or a clear pine or an engineered wood product. Since the wall and the new wood would be painted, the homeowners ask what difference it would make if they substituted something else. No one would necessarily know except the homeowners themselves. Jim contemplates some future homeowners like themselves encountering a "deceptive" substitution in a restoration project, expressing concern that his own agency might be construed negatively as he and his wife have characterized the agency of previous owners themselves. In avoiding a kind of displaced shame over a potentially inappropriate restoration decision, Jim and Karen are also expressing a philosophical position about honestly using "original" materials in restoration to satisfy their own identity aspirations and wins them respect from among other purist homeowners in their local preservation community. It is in this context that the sentiments of stewardship are fully expressed.

For practical purposes, homeowners' practices tend to focus on specific features—such as breakfronts and fireplaces—or on the "rooms" that make up the house. Rooms and room elements are conceived as parts of a whole, the collection of rooms comprising the stylistic totality of the house. The assemblages of items in rooms that make up the house are made to cohere around the aesthetic framework based on the historical style. As they are integrated into the material expression of the aesthetic ideal, they constitute the homeowner's identity and lifestyle, and make the restoration a moral project. While the basic strategy in restoration work is to include as many original forms and materials as possible, a more challenging issue concerns what kinds of substitutes are permissible when original elements are incomplete or missing. This dilemma forces homeowners to imagine what might have been and to justify alternatives through various kinds of research as part of their restoration practices. The prospect of living in an early twentieth-century house, restored to its exact original conditions, might be a bit daunting for life in the twenty-first century if it means living a period appropriate lifestyle. Thus, some kinds of new materials and products, "modernizations," are necessary to make contemporary life possible, comfortable, and safe. This raises the question of aesthetic integrity and the extent to which a "purity" of style—a rigid adherence to period integrity—can be achieved in restoring the older house.

Restoration practices … paying for services …

For many homeowners, the costs involved in restoration work dictate the kinds of craftsman they can hire—like Jim and Karen's bargain floor refinishers. Margaret and Suzanne from Alhambra said they did most of the stripping and painting inside their house, but when it came time to restore the built-in dining room cabinet, they decided that since it was a central feature they would spring for a specialist. Margaret says:

> We had a budget for home improvement, but we decided that we wanted to invest it in the built-in to make it look as best as it could. And we knew that we couldn't

handle the dentil molding and all the tricky stuff. And because we were so in love with it we wanted it to look, you know, primo. So, we hired a guy, happily, who knew what he was doing and he did a fabulous job, we think. And he—even he was amazed. We thought it was going to be mahogany under there. It was not painted but was real dark. It had that kind of tacky, sticky look—kind of thick, kind of alligator … To our astonishment he took it off and that was what was under there.

Others, most often women, describe a kind of close working relationship, one of trust, they develop with some contractors or craftsmen they hire for remodeling projects. A few homeowners hire contractors to take complete charge and make most of the decisions, especially if the house is in very poor condition and is unlivable. When Carol bought her 1929 Mediterranean Revival bungalow in Monrovia, the place was full of trash and had not been maintained; neither the plumbing nor heater worked. Not knowing where or how to start, she hired a contractor who was married to a friend of hers and who, she says, felt sorry for her. He proposed charging her a 10 percent markup for any craftsmen he brought in, he wouldn't charge her anything on people she brought in, but he would coordinate everything. "He was awesome. He just felt so sorry for me."

Other homeowners develop working relations with their contractors using them as expert consultants, even partners, in realizing the restoration goal. Julie, also from Monrovia, referred to hers as "my carpenter guy" who advised her and worked on her window restoration project:

> My carpenter guy, you know, he is just my right hand—he is fabulous. We were worried about how big the center section should be. Should we mimic those windows, should we mimic something else? Should the middle be fixed, should it be double hung, should we have casement? So we sweated over these details—how wide, how short?

Paula from Pasadena also described her "fabulous guy" in glowing terms:

> Again, my master magicians came in here and literally on a Friday morning I had a wall air conditioner that even cut into part of the baseboard, and Friday afternoon, it was gone. It was like that. They restored the cut into the base board, they re-plastered and did the sand coat. And they took off all the horrible aluminum hardware.

These craftsmen are highly valued and often recommended to other preservation homeowners simply because of their specialized knowledge of, and experience in, remodeling old houses and helping homeowners retain or restore their original qualities. But the intimacy that a homeowner develops with a craftsman in making design decisions and restoring a home might also lead, on rare occasions, to romantic involvement. This happened to an Ontario homeowner who was restoring one of her family's houses in which she was living. After divorcing her first husband, she and her second husband, the contractor, are now restoring the Craftsman house where she grew up.

Of course, not all contractors and craftsmen are necessarily sympathetic. On the occasions when homeowners want to install modern conveniences, but effectively hide them from view, the challenge of finding sympathetic contractors looms large. Replumbing and rewiring an old house can be done without revealing the new infrastructure, but heating and air conditioning systems sometimes pose problems. Laura and Steve complained that contractors are really stubborn and have only one way of thinking about how to do a job. Steve said, "All the air conditioner contractors told us, you're going to have to put in boxes to get the ductwork up there and I figured out a way to get pretty much what we wanted without the boxes." Laura from Pasadena described the process:

> Yeah, [Steve] spent a weekend crawling through the attic trying to find a way to get the ductwork through so that we didn't have to break open any walls ... And the contractors also thought we were insane because we were not allowing them to take out our 1918 gravity furnace and put in a brand new thing. And we're like, no, that furnace is working beautifully and it's so quiet ... We only want air conditioning, and we will find a way to do it without impacting the house.

Laura was able to locate old floor grates at the salvage yard and had them brass plated and installed in the ceiling to use as air conditioning vents. The contractors had insisted that they would "have to use the plastic directional flow things or else it won't work," to which Laura responded, "No, I'm sorry, I will not have those plastic things in my ceiling. You really [have to ask yourself] how do you make this look like the rest of the house?" The moral project asks, in brief, how does one stay faithful to and retain the original integrity of the old house?

Restoration practices ... special elements and entire rooms

Preservationist homeowners employ a variety of restoration strategies to resurrect and revalue the "original" features of their homes. The restoration of the fireplace seems to be a benign, but style-defining element because it operates as a centerpiece of the living room and is often a subject of much serious (aesthetic and functional) consideration. But a functioning fireplace, at least in Southern California, is not a necessity for everyday life. Rather it operates more on a symbolic-aesthetic level, as a contemplative focal point, and thus permits a wider range of less than critical choices and interpretations. On the other hand, kitchens and bathrooms are necessary to everyday life and homeowner decisions about them seem more agonized. It may be quite easy to strip the paint off wood moldings and built-in cabinets to restore the original finish, but living with a truly "period appropriate" icebox or wood-burning stove instead of modern appliances is another matter.

For the self-identified purists who adhere to a strict interpretation of what should be left intact and what can be altered to meet contemporary lifestyle needs, these rooms pose special challenges. Still, most purist homeowners consent to having modern appliances even if it betrays their commitment to totally restoring the original conditions of their houses. In regards to fireplaces and kitchens, it is not so much the material qualities that homeowners possess or covet, but the philosophical arguments, the logic of period appropriate restoration strategies they marshal to explain and justify their practices, and critique others. By the same token, those who claim they are "not purists" are eager to demonstrate more flexible modes of restoration stressing practicality, comfort, and an opposition to the rigidity of what they call a "museum" setting implied by strict adherence to purist strategies.

Lucky is the homeowner whose fireplace is still intact because no decisions about how much, or how, to restore it need be made. Patricia said her favorite thing in her 1912 Craftsman bungalow in Alhambra is the fireplace. "I always wanted a fireplace in my home—always. It's the tile work right above—the Batchelder tiles—the little unique ones right above. I like how it wasn't just a giant slab of something. It's so well done." More often, fireplaces have suffered from modernizations of many different sorts that make restoration a challenge. While Karen and Jim in Pasadena had seen the previous owner trying to modernize their fireplace, and quickly took steps to strip away the false façade, others have had to investigate the authenticity of the remaining materials. Paula and her boyfriend spent one evening pulling off the brick that the previous owners had applied to cover the original fireplace and discovered tile underneath—but was the tile original to the house?

> I want to do something with the fireplace, so I seek out my sources. It could be looking at bungalow fireplaces, or going down to Mission Tile and talking to their designers, which is what I did. Then we came up with a plan for the fireplace. That also involves finding out what kind of tile it is. Joe [her boyfriend] is convinced that this is from the '60s. And I found out that it is actually from the '20s and '30s ... So it's basically just starting with what's there now, *what would have been there*, and just knowing where to go to get the information. So, it's a combination of books, talking to people I know who have either done the restoration before or have seen this stuff. [Emphasis added]

Julie discovered that the firebox for her fireplace had been rebuilt after a major earthquake, and the original tiles had been lost in the remodel. She decided to put in antique tiles when restoring the façade, but soon heard complaints from purists who argue for a more period appropriate approach she disparagingly called a "museum" ideal.

> [The purists] say, "Those tiles, they look like they are 1906, 1910, this house is 1920s. You should have had Batchelder tiles." I said, "I don't like Batchelder tiles." That's just blasphemy to them. I said, "I really wanted a Victorian house, but I couldn't afford one, so I have lovely tile that I just love and enjoy all the time."

Carol discovered that her mantle's distinguishing plaster features had been painted over, and were only revealed when the contractor cleaned off the chalky white coating. Still, the bas-relief scene of the San Gabriel Mission was hard to see, so she hired a movie studio scenery painter to recreate the landscape in soft colors. Carol did not seem particularly concerned about how it had looked originally, and there was no way to know, but she was pleased with the creative results of the artist she hired.

Of course, sometimes the homeowners discover that their house once had a fireplace, but no longer does. Does the preservation homeowner attempt to rebuild the fireplace, or accept that as part of the history of the house? Laura recounts the story of her missing fireplace in her 1909 Transitional Victorian–Craftsman in Pasadena:

> The biggest mystery about the house was probably the second owner, because we know there used to be a fireplace right here. Because when we were first looking at this house, I had a rule that I would not buy a house without a fireplace. When we were looking we noticed that on the outside of the house there was a seam in the siding on both sides of that window. And we dug in the ground and found that there was a foundation, a concrete foundation out there and we supposed that there used to be a fireplace. And we talked to owners who lived next door whose parents had lived there and there was a story about a lady who lived in here and … removed the fireplace because it was too dirty.

In part, because the fireplace was removed within the first ten years of its original construction, Laura says she can accept the removal of the fireplace as part of the "original" conditions of the house. Fred and Bart, however, complained that the fireplace in their 1912 Craftsman, in Pasadena, had been remodeled in the 1950s when the Craftsman style was not very popular. "And in the living room, the entire fireplace surround has been … Its very boxy and white and sort of 'federalist' style, you might call it. It's not, you know, an oak mantle with corbels and simple craftsman details." Although the fireplace was not on the top of their list for remodeling projects, given enough time, it might have become a more pressing concern.

Bathrooms and kitchens, by contrast, get a lot of attention because homeowners depend on their functionality to play central roles in everyday life; appliances and technological furnishings must operate correctly and predictably. Of all the spaces in the house, restoring the kitchen provokes the greatest tensions between the desire to retain, conserve, and restore the "original" features and the demands of modern living. As a rule, self-declared purists say they leave the remodeling of the kitchen to last because it's expensive and its functionality requires many difficult choices. In a way, the challenge of remodeling the kitchen requires not only making decisions about what original features to keep and what modern elements to include, but also requires developing a narrative that justifies the choices. Here is where the material integrity of the house comes under the greatest threat. And, while there are also similar challenges in bathroom remodels, retaining leaking old plumbing is not considered an option when many new reproductions are available. There is more noticeable anxiety about the kitchen remodel, however, as homeowners try to decide what to do. Although minimal is usually best, most kitchens have already suffered from extensive prior remodeling that makes leaving the kitchen alone almost impossible. Furthermore, many families who have children or entertain a lot say that modern conveniences are essential, and even some purist homeowners view the kitchen as an exception in terms of function and subject to all the modern conveniences.

Many preservation homeowners say that original kitchens are very rare because they are likely to have been remodeled multiple times and badly. Karen and Jim haven't begun to tackle their kitchen but they do have original moldings, and original cabinets, even if cheap replacement doors were attached to them. Jim explained:

> We don't know what we're really going to do … We've been talking about it for three years and haven't decided. I'm really adamant that we save what we can because it's just very hard to find, you know, and it's really exciting when I see an original kitchen. Like some of them are redone with pseudo Craftsman style, but I really want to save what we can.

The kitchen in Dan's 1913 Craftsman bungalow, in Pasadena, is about as close to the original as possible, which he is trying to furnish with vintage appliances.

> And here is the kitchen … no toe space except at the sink. Yeah, one of the few original kitchens in [the neighborhood]! The tile was probably done in the '40s— and it has a stove and original icebox. It's not running but it works, it gets cold. I just don't run it because it's so noisy. I saw an ad [for the stove] in the recycler—so I'm just the second owner. They wanted $900 for it, so I thought it was bargain. But boy is it heavy. And then the refrigerator—I really wanted one of the old GEs. A friend … called me one day, he lives over in Glendale and … There was a garage sale across the street and I came over and got this one for sale for $50 … And you know what was wrong with it? The fan belt was broken. It still has the original charge of refrigerant in it from 1933.

Dan explained that his Frigidaire was made by General Motors, and that it was built the same way as it built automobile bodies at that time. It has a wooden body to which metal panels had been attached and, he said, it weighed a ton. But, he explained, the temperatures are accurate to this day.

On the other hand, when Dan was asked about his approach to restoration he indicated he only had to please himself, which implied that others might not be able to retain the same level of originality.

> Since I live here by myself, I can pretty much do what I want. The one thing I feel is that the more original the house is, the more it will be with me in the long run. For example, my approach to the kitchen, you noticed that I have another refrigerator back there. So, that's the refrigerator I use. And the old stove, it works fine for just me and it's probably worth three times what I paid for it.

Single women homeowners, however, seem to take a different view and categorically state they want a kitchen that performs; this may be because they cook or entertain more than single men. In fact, a well functioning kitchen is critical to Julie from Monrovia's idea of a kitchen because, she says, she cooks and bakes every day. For this one project she hired a kitchen designer to help her with the project. She had saved pictures of things she liked and kept them in a notebook:

> I wanted a dual fuel range and you are not going to find that in a vintage looking stove, and so I did enough in there so the appliances that you see are my stove and refrigerator—you don't see the dishwasher … those are kind of hidden. But they are very functional and it didn't bother me. I just picked stuff that would

go together that is not trendy, because I will not live long enough—I will not redo the kitchen, ever. This is it. And I didn't want something that says, oh, you have old cabinets and you redid your kitchen in the 1980s … in the 90s, oh, you have stainless steel fronts, you did it in the 2000s. So I tried to pick stuff that was timeless that do not show when I did it.

Preservation homeowners often argue that remodeling a kitchen to contemporary tastes will eventually make it look dated, which should be avoided at all costs, and that restoring or recreating an original kitchen allows the homeowner to escape this temporal and moral dilemma. Laura argued that a kitchen should really be designed to match the original conditions of the house because it is so easy to tell that a kitchen was remodeled later and to date the remodel to a specific era:

Now everybody laughs at 1970s kitchens, but in the '70s they were thought to be a good thing. And in 10 years everyone is going to be laughing at '90's kitchens because there is a sensibility with each decade that people think of as great. But with respect to the long expanse of time its just absurd. So, if you go back and look at the way the house was originally, and try to figure that out and do what's not appropriate to the '90s, but what's appropriate to 1910, then you end up with something that might be able to stand 20 years of time because it is not going to look out of place in 20 years.

In Laura's case, the house itself must contain the clues for remodeling, and because there is so little original material left, she doesn't have much to go on. The previous owners remodeled again, just before she bought the house; "I wished they had just left it be, I would have much rather have dealt with that in whatever state it was in on my own …" And she admits, "I am fearful of the compromises we will have to face with the kitchen because … I have no particular need for fancy modern appliances."

Paula has two very different kitchens in the two houses she owns in Pasadena, the 1922 Pueblo Revival bungalow, where she currently lives, and the 1917 Craftsman to which she is moving. The contrasts she makes illustrate common notions among preservationists with regard to kitchen remodels. In the 1922 Pueblo Revival house very little has been changed over the years. A wall separates the kitchen proper from the service porch, a space that according to Paula was originally screened, but later enclosed during the 1940s, eventually with jalousie windows, which she just replaced with a new fixed window. The kitchen has been painted and the original floor was replaced with new ceramic tile.

I was going to do linoleum (on the floor), but now that we will be renting, I'm not going to. All the cabinets are original and I don't believe the counter tops are, but the size and configuration is original. And no toe-kicks, so … I mean, I love it, it's great. Original kitchens in this neighborhood are hard to come by. I'm sure the trim is original but the tile is not … . This is a California cooler—I replaced these, but I still have some of the original shelves and of course it opens at the bottom— there is still a little vent out there so when the fan blows, and when its really cold during the winter, I can really feel the air.

When asked how she decided what she would keep, Paula admitted she would likely keep changes to the house made in the 1940s, but for changes made in the 1970s she said:

> It depends ... The fact that this was a screened porch, it's not really practical to take it back to a screened porch. And it's also—I think that it makes that cutoff— that in the '40s we started to have refrigerators. We don't have iceboxes any more. Maybe we need room for laundry and so this is very typical to enclose their screened porch to make it into a service porch. So, it's a comment on the evolution of the house during that period, so that is why I would keep it. I wouldn't—I mean I would replace the door, which I still might, 'cause this is not an original door. But I would keep it a service porch—plus, it was also done by the original owners. But, like, people might come in here and say they're going to gut the kitchen. They're going to knock out that wall. Like they could if it's not a load-bearing wall and

Figure 12 Remodeled kitchen with original cabinets without toe-kick and new cabinets to the right with a toe-kick. Credit Denise Lawrence-Zúñiga.

people might redo this and make it into one big kitchen. I would never touch the walls in here. Out there is a portable dishwasher and I roll it in here and I plug it in here—I don't have to lose cabinets.

Here, Paula outlines a number of remodeling strategies including something about bending to the practicalities of contemporary lifestyles that depend on appliances and laundering requirements, but where a dishwasher is concerned, she argues a portable machine is preferable in order not to "lose" the cabinets. She also reveals her preference for changes in the service porch made by the original owners of the house. It is as if only the original owners had the right to make those changes. Her position acknowledges that the house continued in their possession for fifty years and bore indelible primary marks of their agency and theirs alone, meaning that successive owners had less of a right to make changes.

Paula argues that it is not difficult to live in the 1922 house. "Its really a balance between, do I want to live the way we live now, or do I want to try to fit into the way they lived then? To me it hasn't been a hard fit, with the exception of the no toe-kick in the kitchen—it really does get to the back."

Although the only change she says she would make is the addition of the toe-kick, it is clear she is not completely comfortable with the limitations her 1922 kitchen place on her lifestyle and looks forward to moving to the new house, the 1917 Craftsman, where the kitchen was completely "redone."

> Considering the amount of cooking we do, the amount of parties we have, it's not a good kitchen. But I won't touch it. I wouldn't touch it. And the new house has a totally redone kitchen and it is not what I would have chosen, and I don't care because finally we have the counter space we need and we're going to be able to put a table in there. Because they have knocked out the wall to the service porch, so it's finally this eat-in kitchen that is nice. *As long as I am not the one doing it.* [Emphasis added]

Paula describes that the new house does not come with appliances, so she and her boyfriend have bought a new professional stove and stainless steel appliances because they feel they "have more latitude." They are not constrained by the original kitchen because someone else has already destroyed the original kitchen.

> But we did think about getting an old stove … There are limits to what I'll do—I mean I love the look of the 1920s stoves, that's a little too much for me. That's a little frightening to me—I need something that's a little more comfortable. And most people in this area, no matter what year their house was built, will settle on a '40s or '50s stove. I think it's kind of interesting … I think maybe it gives you this space you need—the modern—the ability to really control the thermostat and everything.

So, for all the commitment to preserving the past, Paula betrays that love of the original with her love of modern technology.

> So I am really excited about the French door refrigerator with the pull out door freezer at the bottom—which is where I think all freezers belong because you don't

use them all that often. And I can't wait for my Bosch dishwasher with the hidden controls and that Thermador stove. I told [my boyfriend] I am going to sleep on it because I am so excited. ... and I can hardly wait to start doing laundry in my big LG washing machine and dryer. But it really depends. If that kitchen had been original, you'd better believe that I wouldn't have touched it. But when they've already gutted it and redone it, then it gives me latitude—permission—especially in the kitchen.

Reliving the past ... constructing a purist identity and lifestyle

Preservation homeowners are critically aware of the needs for comfort and convenience in the restored house—much of the discourse in the preservation community revolves around negotiating the demands of everyday life with a material context that is not always that forgiving. As Tom remarks: "Comfort is definitely an issue—we do have double hung windows and they are drafty as heck ... [But] I'm living in California, so it's not a life and death situation. If that were back east, it would be different."

But many of the homeowners are more than willing to retain original features—a single bathroom in its original configuration—in part because it pleases them to be able to live in the home as the original occupants lived. A number of homeowners stress this in their explanations of their restoration strategies. Julie says:

I would like to think that if Ott [the original owner] walked into this house today, he wouldn't be too shocked at what he sees. The footprint is the same. There are bookcases where he had bookcases, and it's ironic that in the kitchen, there really—it hadn't been upgraded since the '20s ... So, when they took all the walls out and all the ceilings out and all the floor out, and you looked at how the room was framed up, we wound up putting everything back exactly where it was when it was built. The stove is exactly where the stove was, the washing machine is exactly where the scrub sink was. Where you sit to eat was exactly where the nook was and we didn't know that until we tore the walls out. But it worked.

Paula also describes her vision for restoring of the house in terms of the original occupants' presumed point of view:

The bathroom cabinet was just a desire for what needed to be there. I want to be able to walk around to see what this place is supposed to look like. Well, I want to know what the first residents saw when they were in this house—that needed to be restored. I love the idea of uncovering something and finding it there. I was so excited when I uncovered this [the fireplace]. I mean I knew that that brick was going to be gone as soon as I walked in the door ... And I discovered ... what it is supposed to be with this probably original paint color. So, we've got this nice plaster surround and this tile inset and I just thought that's fascinating.

Paula's attribution of a normative quality to the state of the house—that its original condition is morally superior to any other—informs the preservationist's motivation for restoration, and operates in the co-production of identity and lifestyle. But restoration presents other kinds of challenges:

> When I see something like this, I get really excited because it's just another piece of the personality of the house. But the problem is that there is a kind of paralysis, which is why it is taking me so long ... Do I try to find out what was on the fireplace? Do I have a responsibility to restore it back to the way it was originally? Is it ok if I choose something indicative of what is appropriate to the period? What the heck do I do? So, that's what happens—and it sits there for a year.

While Paula's previous restoration strategy for the enclosed service porch had been informed by the remodels of the previous owners who had lived in the house so long, and imparted legitimacy to the material changes they made, resolving the dilemma of the fireplace turned her attention to the original designer's intentions.

Laura and her husband, Steve, explain their strategy for what turned out to be a complicated restoration project in their 1909 Transitional Victorian–Craftsman house.

> It started out upstairs because I was painting the hallway and the upstairs hallway is the only room in the house that does not have a picture rail. And we noticed that as we were first painting it white—the first time—that there was this little funny bump under the millions of layers of wallpaper and paint that we interpreted at the time to be the scar of a picture rail that had been removed.

Laura had deduced the presence of the picture rail from the fact that all the other rooms in the house had had one. They planned to get dies cut to reproduce the shape of the picture rail, but in the meantime Laura discovered the little "bump" on the wall had actually been a one-inch strip of wallpaper border that had been painted over multiple times. Once she had scraped a piece of it off, she discovered a chevron pattern.

> So then we went on a search of finding 1-inch period reproduction wallpaper border. And it took a while and we found a pattern that works. It had been painted over so many times that it left a lump that we didn't know what it was. So, frankly, I really wanted another piece of picture rail because I've got pictures in the hall and I love hanging my pictures off the picture rail. And if it was just up to me, I would have just gone ahead and had the wood cut and put up there, but I couldn't do that because now I knew that there had been a wallpaper border there and so, that's what we had to do. *Because we let the house tell us what to do.* [Emphasis added]

Here it is not so much the previous occupants' repainting of the hall that informs, but rather to discover the original designer's intent. Laura patiently resolves the question of designer intent by ascribing agency to the house—the house is definitive, it tells them what to do.

These imaginary conceptualizations are not just flights of fantasy created by bored homeowners, but serve their need to visualize the embodiment of the three-dimensional material form of the house in an historically consistent fashion. The imaginary construct facilitates the import of facts gleaned from reading and researching texts on

Figure 13 Restored wallpaper "bump," a running line aligned with door-tops.
Credit: Denise Lawrence-Zúñiga.

architectural styles, city archives, family genealogies and narratives, and local history, which are assembled within and around the materiality of home. It makes living in the house the embodiment of a kind of experiential history, of time travel, that becomes deeper and richer as more information and experiences, are assembled and enacted in its spaces. Paula says:

> I wish the house had a second bathroom. But, again, I feel like it has been like this for 80 years. Is it so bad that we maybe try to take cues for how people lived in the past? I mean, just because we live differently now, doesn't mean that there aren't some good points about the way people used to live … The bungalow was designed so that mothers didn't have to go up 17 flights of stairs and so older people could easily get around. It was all about easy living and efficiencies.

The imagined past is a dynamic construct open to amendment and correction, as well as addition of knowledge of the past and future activity. It is realized by discovering the agency of others who have made their imprint on the house and by the current owners who legitimately invest their own agency—not as a fleeting or temporary fad, a style—but a lasting contribution to the durable material record that someone in the future may inhabit and revere.

The reconstructed view that integrates historical information about the style, period or other houses also serves preservation homeowners in articulating explanations for their restoration choices. Paula explained that when she replaced modern light switches with reproduction push button switches, she had to know that they had likely been there originally. "When I do stuff like that, I want to make sure that it can be justified—is this something that could have been here in 1922?" she says. "And the answer is yes. So, I wasn't going to put it in if it couldn't be justified." The construction of the fantasy is also central to the homeowners' sense of stewardship in preserving the house. When asked if she thought of herself as a caretaker or steward of the house, Julie replied:

> Maybe [my role is] to leave it better than I found it. I always think of the caretaker just kind of keeps it as it is. But the next family that moves in here after me—that they won't bulldoze it and put a three-story, 5,000 square foot house on this lot, will enjoy this house and realize that the children in this house have been people who lived in it and loved it. Sometimes I sit in the bathroom and think how many people—how many guys—have stood in front of this same mirror, the same medicine chest, and shaved in there all these years. You think of all the people who have come through this house and lived in it and enjoyed it just like I have. And its like—not that there are ghosts here—but it is like it—all these lives that have come through this house give it this lived-in feeling.

The stewardship role links the agency of past occupants and the designer/builder to the embodiment of enduring material qualities of the house. Homeowners who embrace stewardship link their own agency to the long lineage of prior agents by engaging in restoration practices. They see their responsibility as one of protection—Paula describes this responsibility as a negotiation between what she needs, how much she actually knows about the original qualities of the house, and what is possible or practical:

And so, it's like what is my responsibility to this house? You know, and I mean I think that anyone who lives in an old neighborhood or an old house, I think we have a responsibility to take care of these houses and make sure they last another 80 years. And, we're kind of stewards of them and I think that we have to ... I mean I am saying this to someone [the interviewer] who is trying to make her house fit her lifestyle ... I mean that's the other thing, what changes would I make to this house if I were going to make this house fit my lifestyle? I think that maybe the only things I would do ... its really hard because preservation is such a part of how I feel that it is kind of difficult to separate what I would really want if I was designing it from the ground up. You know what I am saying?

As Paula clearly indicates, these practices are transformative. As preservationists become increasingly involved in restoring the material qualities of their homes, they develop identities and lifestyles built around historic preservation and the restoration of home. As Laura summed it up:

In many respects, I consider this house to be the thing that focuses me with everything that we do. Looking at the house itself and the way other things in the house have been done and trying to draw from that. Regardless, in some respects, of what I would like to do myself, you have to listen to the house.

Eventually, "the house becomes a part of you" (Patricia, Alhambra), a realization that many homeowners come to when their remodeling projects require a bending of lifestyles to fit the house.

Not such purist remodelers ...

While the most dedicated of preservation homeowners express a deep commitment to the ideal of preserving original features, many others explain they have used more creative or idiosyncratic interpretations in restoring their houses. Like purists, these homeowners also claim to be stewards of historic architecture and profess adherence to similar goals in restoration. Their descriptions of their remodeling practices, however, admit the easy incorporation of more modern conveniences and openly suggest taking more creative license in formal interpretations of style in their projects. Some of their creativity is in response to the amount of destruction of original features in prior remodels and modernizations, while other ideas seem to spring from the homeowner's own aesthetic desires. Like purists, however, the imagination plays a powerful role in developing a remodeling strategy that coheres with the original qualities of the house.

Jack and Nellie owned a 1921 Colonial Revival bungalow in Ontario, which they restored before moving to the 1908 Craftsman in Claremont, they are currently remodeling. In both houses they express a commitment to the ideal of the original, "We want to restore those [crown moldings] so they look exactly like they did when the house was built." But later, Jack describes a more creative interpretation by imagining the past in their remodeling efforts:

We like restoring things back to what they *might have been*. We like mixing in some modern things with it once in a while like … that sculpture piece they just delivered. So what we find beautiful is the craftsmanship in different pieces … We redid the master bathroom in our Ontario house. When people walked in, they absolutely thought the bathroom had been there since the house had been built, because the materials and things that we used were traditional … We made the house look like how it *would probably have looked* when it was built. [Emphasis added].

Tom also takes some creative license in restoring his 1908 Craftsman in Monrovia. He has been remodeling his kitchen using his own workshop to craft new cabinets and adding modern touches. He says that nothing looks out of place in his purist neighbor's Victorian kitchen; even the microwave is completely hidden behind doors. He draws the distinction in describing "compromises" in his kitchen remodel strategy:

Me, I'm compromising. Can lights—they're efficient, but there are some things that are going to be old looking. The lighting is not good—we used to have fluorescent lights, which were marvelous but they looked awful. So, there are going to be those concessions. On the other hand, the cabinets will have an older look to them and we're going to put linoleum on the floor, which is an old, old, product—you can go back to the mid 1800s. So, there's a compromise of things. We're going to put a fan in the center …, but strictly speaking it doesn't belong. And if you want to be strictly speaking, cabinets in the kitchen were painted. Cabinets in the kitchen sat on the countertop … So, how purist do you want to be? … We live differently. Our lifestyle is different from 100 years ago—I don't think it is realistic to try and live that way. I mean, you can, certainly.

Tom's approach to remodeling his living and dining rooms, and kitchen, seem to be all of one piece. Rather than try to exactly restore what was originally in his house, he has designed solutions to lifestyle needs that draw on his knowledge of other Craftsman houses, including his previous home in Alhambra.

With his extensive carpentry work, Tom is investing his own agency in creatively interpreting Craftsman-style materiality. He built new bookcases for either side of his fireplace, even though there hadn't been any there originally, and, rather than recreate the missing built-in cabinet in the dining room, he built a new cabinet in the kitchen to hold dishes and linens.

I'm putting wainscoting in the kitchen to match what's in the dining room to improve the *sense of connectedness* between these 2 rooms—so that's maximizing (change) but it's within the scope of what we are trying to accomplish. So *we want people to walk into these rooms and feel* like these are old rooms, but not antique completely. And a lot of that concession is also going to come about as a result of age. For example, these lights are very atmospheric, and they are old fixtures and its very nice but it's not that easy to read by. [Emphasis added]

Tom's remodeling takes creative license in interpreting the Craftsman aesthetic to self-consciously produce an effect—"the sense of connectedness"—he intends others to feel. Although the strategy borrows from preserving the original aspects of the house, it is not a literal interpretation of the original. In fact, Tom says that he does not feel constrained by the original if lifestyle needs and practical concerns, coupled with stylistically informed alternatives, are available and possible to produce. He also criticizes the purists for voluntarily limiting themselves. "I think they deny themselves things thinking that it's better. And I'm not sure that it's really necessary to go to that extreme. But again, we all have our own tastes, our own colors that we like."

Elderly homeowners have also had significant experiences remodeling their houses, especially when they have lived forty years or more in the same place. All three homeowners had remodeled their kitchens—or upgraded them and even modernized them. Sam and Teresa had bought their 1936 French provincial-style house in Riverside in 1964, and began remodeling the kitchen in 1968. Sam taught industrial arts at the local high school, and had a shop in his garage. He built all new oak cabinets for the kitchen and modernized the appliances and plan; later he remodeled other interior features. Other elderly homeowners had contractors install modern kitchens, for the time, to accommodate their growing families. None of these remodels was particularly style or period appropriate in the sense that preservation homeowners mean, and at least one couple regards their kitchen as dated and needing another makeover. One of the differences between the elderly homeowners and the others interviewed in this study is the generational difference. As mentioned earlier, the couples are all over seventy-five, and have lived in their houses for at least four decades, and remodeled long before they were admittedly much concerned with historic preservation issues. If they were not currently living in their houses, the next owners might be already condemning them for "inappropriate remodeling."

Family-connected homeowners …

If homeowners, who restore the original qualities of houses with which they have had no previous connection, develop a kind of attachment and identification as described above, what kinds of relationships do the three family-connected homeowners develop with their houses? Are they equally devoted to restoring original features? Do they construct the same kind of purist identities and preservationist lifestyles? As it turns out, unlike the other two homeowners, Stan's strategy for restoring his family's 1887 Victorian is consistent with the purists described earlier—to return as much of the house as possible to its original condition. Stan uses photographs, his own memories of the house and talking to his grandmother to understand how the interior spaces were originally configured and furnished. He explains that his grandmother had remodeled during the 1930s so she could rent out parts of the house and he has been trying to reverse those changes. He notes that the original doors into the parlor were removed and replaced with "an arch there and a double arch over there … And, yes, it opened these rooms up, but it was so antithetical to any Victorian style." Stan also had reproduction moldings made to match the doors where the originals had been removed, put doors back where they had been when the house was built, stripped the six to eight layers of paint from the wood trim, and restored windows and doors that remained.

In many ways, Stan's approach is identical to other preservation homeowners, but he is able to explain more fully why his grandmother had made the changes she did. "I asked her one time about the woodwork, and she said, 'Well, you know, I'm sorry now that we painted the woodwork, but at the time it seemed like a good thing to do.'" He reports that "she said she got tired of opening and closing doors" to explain the arched open doorways. Stan says his house has two kitchens, one added in 1914 and another in 1924, which his grandmother later retiled, but otherwise both are in "original" condition. When asked about acquiring modern conveniences for the old kitchens, Stan said:

> Well, appliances—a kitchen full of things that are really handy and allegedly make your life easier. And then you think—do I really need this stuff? A television in every room? Every appliance here has a sense of how people lived 100 years ago. Part of my view is knowing, or not wanting to go back to that, but thinking that their quality of life was better than ours and they did it without many of the things we have. I don't have air conditioning. And if it gets too warm upstairs I use a box air conditioner. And I don't feel deprived … I don't feel deprived because I don't have it.

Stan acknowledges that he does have a washing machine and dryer, but as a single male, he is realistic about his lifestyle preferences:

> You know, I'm only going to do this for myself. But I'm not willing to subject anyone else to it necessarily. So, if somebody else brought strong objections to being that authentic, then I don't think I would try to make it that authentic. But, if you are by yourself, like I am, I am not really impacting anyone else. Nor am I asking anyone else to really sacrifice. So, I would tend to be as authentic as possible, knowing that when there is a husband or spouse or children, you have to temper that with reality. You know, the scrub board on Monday just doesn't do it."

As with the other purist homeowners, Stan is committed to returning his house to its former glory, and is hoping to open a bed and breakfast business. The other two family-connected homeowners, however, have taken considerably more license with their remodeling efforts. George, for example, has pursued the restoration of his 1887 Queen Anne in Ontario back to its original exterior configuration by removing a mansard roof, shown in photographs, which had been attached during the 1920s.

He has also stripped paint and wallpaper, and acquired reproduction and antique hardware, lighting and bathroom fixtures to reproduce the Victorian character of his house. He imagines passing the family home on to his high-school aged daughter who is just now developing an interest in the home. On the other hand, he hired an architect to design a large addition to the rear of the house to accommodate a new kitchen below and master bedroom and bathroom above, all with modern amenities but in the Victorian style. "The [original] kitchen was totally dysfunctional—modern appliances are fine and all the conveniences. We'll do a farmer's sink—we'll do what is appropriate, but modern." As part of the project, George installed an elevator next to

Figure 14 Photograph of the Queen Anne before it was remodeled. Courtesy of Richard Delman.

the bedroom so that he and his wife can retire in the house without having to worry about getting up and down the stairs. He said, "I'm not going to change the interior of the old house—I'm going to make this look like the old house as best I can with new construction."

Sarah shares her family's 1909 Craftsman in Ontario with Ben, her contractor husband, and their children. Although she has many fond memories of her parents and of growing up in the house, she also admitted having some difficulties letting go of the past. Moreover, living in and restoring the house with her husband has meant a negotiation of their relationship with reference to her memories and the material agency of the house and its furnishings. As mentioned earlier, the house had come with original Stickley furniture—now quite valuable as antiques.

> But I'd say we've reached a balance of respecting the house or its architecture and not being tied to the fact that it was my family's. I mean I think that its neat that it was my family's and there are things of theirs still here, you know. But things people would put in any home—if it were a new home, you know, old pictures and things like that. But I finally feel like it's really our house.

Figure 15 Queen Anne house restored to its "original" condition. Credit: Denise Lawrence-Zúñiga.

Ben agreed, but added "the tipping point on that was these couches and that chair, because it was like a Stickley museum, really." He described the old Stickley settle, a wood-framed couch, as an uncomfortable piece of living room furniture that was impractical for a family. Once Ben was able to convince Sarah to sell it, they got $11,000 and were able to buy two comfortable leather couches which helped "to dispel the museum quality" of the room.

Ben argues that he is Sarah's "dream-killer" because he is the bearer of bad but realistic news about the value of material items that populate Sarah's memories of the past. She still cannot part with the wooden Stickley basket that held firewood next to the fireplace because "these are some of the things I still can't let go of. We had this wonderful gardener that came in everyday and would light the fire at four o'clock before he went home." To which Ben responds saying Sarah's life is highly romanticized and that he is always "fighting her concepts because [she] has this wistful view." When I suggested that hers were "actual memories," Ben protested. "Oh, but that's where I disagree. These are romanticized and apocryphal memories, you know, that are inflated. I think its sweet and I like it, but I say, that's bulls**t." Ben's vision, which has also become Sarah's, is that the house should be made a "great place" for everyone. He observes:

> I don't think you do any service to the historical quality of the house, or to the life of the house by making it too precious, and too authentic, and too locked in to some abstract notion. And I do think some of these notions are abstract and really, not even accurate, a lot of times. You know, a house has to be lived in—it has to be a dynamic thing, otherwise it becomes a museum. And then what the hell good is that? I hate that.

As a contractor who has restored other Craftsman houses, Ben likens preserved houses to museums, to being "dead—not a living thing," implicitly critiquing the purist approach also seen in other homeowners' comments. Sarah, however, has a slightly different, more personal view:

> And it goes back to our thing with the furniture. It was very important to me to make this house lively and alive again, because of the years when it was tough, you know. My mother died here, my dad died here; it was tough. They were downstairs and it was not a happy place. So, for me it's important to have it be lively and bright and have kids running around ... to make it continue to be something vibrant. Like where it was.

This sentiment, as an alternative to the purist concern with keeping or restoring or "freezing" the house in its original condition, is also found among a few homeowners who have had no family connection to their houses. Sharon and Tony, for example, live in a 1936 "vernacular cottage" built by a Swedish immigrant in Riverside. They have been restoring the house but do not consider themselves "stewards" or caretakers of the property, although they admit to have taken steps to protect the detailing. Tony, who emigrated from Poland, says that after World War II destroyed much of Poland's material culture, his family lived with cheap mass-produced products while the Communists took away the antiques and heirlooms. He says that after 1989, Poles were challenged by the dearth of material culture for recreating a new middle class. Tony argues, "Historic works of art belong in museums. Houses are human spaces that have continuity and must accommodate the warmth of people. Living spaces are not museums, but are for people."

Conclusion

Homeowner narratives reveal a variety of strategies for remodeling and restoring an older house rather one singular approach. Clearly there are some generational differences between older homeowners over seventy-five years who have lived in their homes forty years or more, and younger homebuyers. While energy for tackling big projects may be one of the more obvious differences, a passion for restoration based on an experiential historical understanding of home seems central. But even those who are committed to restoring the original features of their houses show variations in the way the "original" is interpreted. Homeowners mention making mistakes as they learn what is important to salvage and experiment with restoration techniques. Others mention the challenges involved in balancing the costs of resources needed to produce "ideal" outcomes. Homeowners' narratives describe as much how they resolved aesthetic and technical issues as how the restoration process changed them.

Engaging in restoration practices is at the center of homeowners' construction of preservation identities and lifestyles. These practices include not only a discovery process in which "original" features of the old house are uncovered, but also encounters with the agency of prior designers and occupants through photographs and documents, comparative research, stories, and homeowner imagination as detailed in the previous chapter. The practices of restoring the original qualities of the house also entail "learning to see" and value selected features, as well as defining a concept of a preservation "ideal" that guides the physical transformation of the house and constructs its history. In the variety of restoration ideals homeowners construct, "purists" tend to be the most concerned about resisting modern conveniences or material substitutes, while others are more relaxed. This distinction is illustrated most starkly in their different ideas about restoring the "original" kitchen, but contrasting arguments are also marshaled to explain and justify differences in broader lifestyle preferences. While some homeowners clearly object to the prospect of living in a "museum," preferring comfort or making a house lively as an alternative, purists suggest that submitting to the agency of the house and its history holds more meaning and validity for them.

As mentioned earlier, homeowners seek out and are influenced by like-minded neighbors, community organizations, and professionals from whom they learn about preservation ideals and practices that form part of a larger formalized "cosmology" of the original around which historic preservation is organized. Chapter 4 examines this preservation cosmology as it is constructed and practiced through professional, civic and municipal organizations and entities, and how the cosmology becomes central to discourse about neighborhood and community life.

Historic Preservation as Cosmology: Municipal Regulations and City Dynamics

What is "historic" about an old house? And, of course, why should it matter to anyone but the homeowner? As we have seen in the last two chapters, homeowners often construct quite elaborate experiential histories as their preservation practices uncover previously unknown facts about their houses and their occupants, that they then weave into their own imaginary speculations. The restoration of a home's material qualities brings to life the personalities of the original builders, owners, and subsequent occupants, and some of the major events that occurred there over time. These experiential histories, while constructed around personalities critically involved in a particular home, may lack any significance beyond the neighborhood. Still, what is historically and stylistically "significant" about an old house may be recorded in archives and scholarly texts, entirely independent of the homeowner's experiences. Once discovered, however, those factors lend additional value to the preservationist's experiences and privilege the material products of their preservation efforts. Issues of broader significance may intensify the focus of homeowner attention in the restoration process. Moreover, they also draw homeowners into community activities beyond their own house and neighborhood, and influence homeowners' activities in civic organizations. This chapter explores the larger social arena in which historic preservation ideals are produced and reproduced, and how those values are translated at the municipal level into policies to protect houses and other architectural structures.

The material qualities that houses must have to be considered officially "historic" are defined and codified in an array of national platforms and institutions, and are established in canons beyond the confines of local neighborhoods and communities. A well-established institutional framework, one part of which now incorporates a global vision, is concerned with the preservation of monuments and architecture of "significance." Organized efforts to preserve built features of the environment have well known historical roots in nineteenth-century England and France, and have been embraced during the twentieth-century in the United States, and other nations seeking to preserve their national patrimony. The efforts to institutionalize preservation have been accompanied by a substantial development of scholarly and professional work that supports, engages and critiques the aesthetic values of preservation and other issues related to assessment and impacts. An interdependent network of academics and professionals provide legitimacy to the efforts of local communities and preservationists everywhere. Finally, there is enormous commercial interest in preservation in the form of salvage, production and reproduction of "authentic" historic artifacts, craftsman specialists, contracting, and real estate investment and sales. These actors become de facto defenders of preservation activities as their participation highlights the economic benefits of restoration practices. It is not my purpose here to provide a comprehensive review of historic preservation programs nationally or worldwide, but to highlight features of the institutional context that influence local level historic preservation activities in the five cities under consideration.

The organization of ideas supporting institutionalized preservation activities takes the form of a "cosmology," here a body of interrelated beliefs and practices that encodes particular meanings in the materiality of the built environment. I chose the term cosmology to help situate the extent of shared knowledge among actors at various levels—from household to municipality—and to highlight the boundaries between those who are aware of or knowledgeable about codified historic preservation ideas and those who are not. Homeowners often make reference to these ideas in interviews, even more so than city officials who express themselves in highly technical language. Civic activists also utilize these ideals, in a more impassioned manner than homeowners, as a way to call others to action. Homeowners and activists, with some exceptions, tend to use more popular conceptions rather than the officially codified ideals. The concept of cosmology implies something larger than a simple worldview or set of commonly held ideas. Cosmologies organize deeply held sets of beliefs the consequences of which motivate and empower people to act, and are used to justify or legitimize actions done in their name. Cosmology is a form of power.

Often used to describe metaphysical ideas and sacred propositions that organize religious beliefs and emphasize the cosmic power imaginary actors hold over the universe, cosmology refers to "the body of conceptions that enumerate and classify the phenomena that compose the universe as an ordered whole and the norms and processes that govern it" (Tambiah 1985: 130). The sacred propositions include a society's most revered foundational principles and understandings worthy of perpetuating without serious questioning. Wolf suggests that powerful elites are anchored in a "cultural structure of imaginings" that cannot be explained rationally (1999: 283). These cosmological imaginings generate ideologies that provide elites the support of larger cosmic forces enabling them to act in roles as their agents. Ideologies often reference primordial beginnings as presuppositions for prescribed actions, endowing actors with authority as well as reinforcing the legitimacy of the cosmology with every performance.

The concept of a secular cosmology is useful for describing the framework of historic preservation ideas because, just as religions capitalize on imagined supernatural powers, the material qualities of the built environment are conceived as possessing agency. The material agency of built forms enables history to be "told," imagined, or invented. The doctrine of the "original," which advocates preserving or restoring the original features of old buildings, is core to this set of beliefs and practices. To a large extent, professional preservationists believe that only through retaining a building's original material qualities can history be truly represented and meanings conveyed. The agency attributed to materiality is grounded and justified, however, in the deliberate construction of historical meaning using codified texts to categorize and classify architectural forms and features as "significant" representations of past ideas, events, and personalities. Preservationist cosmology builds a community of interest around a set of aesthetic ideas that motivate believers toward a common preservation end. The preservation ideology that historical imaginings engender are found in the movement's own history, in the philosophical understandings of its nineteenth-century English and French founders, and in the growth of North American professions, institutions, and industry. At the local level, urban planners and local civic organizations, as well as the growing ranks of private planning firms and historic preservation professionals, are the repositories for much of this knowledge.

Beginnings of the "aesthetic community"

To examine the foundational ideas of historic preservation, the concept of material consumption can aid in understanding the core ideas. For the preservation advocate, the materiality of the house is consumed not in a linear process, where the physical qualities are eventually used up to be demolished and replaced, but as a cyclical process of production, consumption, and re-appropriation. The historic materials of house components can be reclaimed, re-appropriated and made serviceable again in a conceivably never-ending cycle (Thompson 1979). Preservationists' narratives certainly distinguish their philosophies of materiality from those they see dominating the modern world and its industrialized mass-production of commodities for the marketplace. As an alternate concept of material reality, the ideology of contemporary preservationists often serves as a form of resistance to modernity, but also re-establishes and restores a philosophical tradition with historical roots in England and France.

Many contemporary preservationists have revived, not always unintentionally, or maintained continuity with the philosophies and inherent contradictions expressed in the nineteenth-century works of John Ruskin and William Morris, who were both influential in the Arts and Crafts Movement. Those authors' anti-modern sentiments critiquing the effects of capitalist industrialization on the human condition provided a philosophical foundation upon which the modern historic preservation movement was founded. Nostalgically mourning for the "world we have lost," they advanced a philosophy of an aestheticized materialistic historicity that rejected machine-made mass-produced artifacts in favor of those hand-crafted with traditional methods and communal labor consistent with medieval guilds (Stanksy 1985). Ruskin, in particular, argued that, beginning with the Renaissance, architecture had lost its traditional capacity to communicate meaning and, in particular, to speak the "truth" (Kaufman 1987: 31; Ruskin 1989). By the nineteenth century, the destructive pressures of capitalism to maximize profit, coupled with the long-term Renaissance impulse to achieve design perfection, threatened to obliterate, or irrevocably alter through reconstruction efforts, the built forms of the past. The potential loss was a call for the establishment of a protective program for the conservation and preservation of "ancient" buildings.

Early preservation arguments about protecting and conserving history focused on selecting aesthetic and material qualities of the built environment to represent and symbolize the past. Ruskin's call for buildings to speak truthfully in terms of their structure and materials, and in the avoidance of machine-made elements, confronted architectural deceptions of the Victorian era that made buildings appear to be something they were not (Ruskin 1989: 35). Ruskin "likens buildings to sentient beings" (Kaufman 1987: 31), and attributes to them the capacity to communicate, to speak the truth, and reveal memories. "For indeed, the greatest glory of a building is not in its stones, nor in its gold. Its glory is in its Age, and in the deep sense of voice-fulness ... even of approval or condemnation, which we feel in walls that have long been washed by the passing waves of humanity" (Ruskin 1989: 186–7). The ascription of social agency to material forms (Gell 1998), the desire of homeowners to hear what stories and past secrets their homes might reveal in the restoration process, continues in these early preservationists' attribution of metaphorical meanings and sentiments to architectural forms. As part of the larger preservation cosmology, early writers and contemporary practitioners shared in a materialist vision of history that they used to validate their metaphysical understanding of architecture.

The earliest preservation leaders were primarily concerned with protecting public rather than domestic structures. Both Ruskin and Morris were instrumental in establishing, in 1877, the first English institution dedicated to the preservation of monuments, the Society for the Preservation of Ancient Buildings. In 1882, the first Ancient Monuments Protection Act was adopted, which was followed by the 1913 Ancient Monuments Act to preserve antiquated buildings of national significance. At about the same time in France, architect Eugène Emmanuelle Viollet-le-Duc was also espousing similar ideas about ancient architecture. Viollet-le-Duc was instrumental in restoring many Gothic cathedrals by engaging in "restoration," rather than the straightforward conservation advocated by Ruskin. The former's preservation work often took creative license with architectural restorations, inventing a more perfect architecture than was originally built, for which he has since been criticized (Tyler 2000: 19).

It should come as no surprise that homeowners' narratives of their own struggle to preserve historic neighborhoods in the face of private capitalist development mirror sentiments and strategies advocated by nineteenth-century preservationists. Should there be a faithful reproduction of the original, or should the restoration improve on the original? Many homeowners pursue self-education in the Arts and Crafts Movement to learn techniques and philosophical positions in order to inform their own practices as well as those they advocate in the civic arena. And, they learn that some of the same forces of modern destruction continue to threaten and undermine preservationists' attempts to salvage the built environment. Preservationists also learn that the cosmology has unintended consequences for those who do not adhere to the same beliefs.

In railing against capitalist industrialization and mechanized production, both Ruskin and Morris expressed deep concern for the dehumanization of laborers on the production line, and found hand crafted work ennobling. According to Ruskin, the absence of human labor makes an artifact worthless (1989: 35); its smooth, perfected machine manufactured surface makes it deceptive, because no work can achieve perfection except "god's work" (McLean 1973). Morris linked the conditions under which an artifact is produced to its artistic qualities arguing that the medieval craftsmen produced objects of greater beauty than do machines (Miele 1995: 76). To save ancient buildings, however, Morris proposed that only the genteel classes, a "small knot of cultivated people," have the time, money, knowledge and understanding to protect monuments (Miele 1995: 75). Herein lies the contradiction that also confronts contemporary preservationists: their romantic attachment to historical materialities, and their original craftsmen seem to produce an aesthetic community among them, the small knot of cultivated people, that excludes the very humble contemporaries who aspire to acquire the modern, machine-produced need-serving goods they reject.

Thorstein Veblen (1989 [1899]) was one of the first to analyze the sociological implications of both Morris's and Ruskin's theories of handmade aesthetic sensibilities as a form of conspicuous consumption. Veblen's concept referred to the grand bourgeoisie—the owners of the means of production who indulged in conspicuous consumption and conspicuous leisure practices. The upper and upper-middle classes could substitute. As members of the English upper-middle (or "leisure") class, their preference for simple "defective," but more costly handicrafts represented an expression of pecuniary taste meant to distance adherents from the mass of consumers (Veblen 1989: 161–2). Ironically, in advocating the abandonment of cheaper machine-made, mass produced commodities for handcrafted ones, Ruskin and Morris advocated putting such items out of reach of the working classes, who had once crafted them by hand, but now made them in factories (McLean 1973: 350). Thus, the materialist vision of history that embraced the consumption of rough hewn handicrafts and the protection of ancient buildings produced an exclusive aesthetic community that rejected the class of workers it had originally nostalgically and idealistically exalted for the crafts they once produced.

U.S. historic preservation movement

Interest in historic preservation in the United States also began during the nineteenth century, although the character of ancient sites and buildings was different. Native American settlements often left behind fewer large permanent monuments in places settled by Europeans, and the colonial presence was not nearly as deep as historical settlements in Europe. Some sporadic attempts to rescue early architectural structures of pre-colonial and colonial significance were attempted in the mid-nineteenth century, but the primary efforts to establish systematic laws to protect colonial monuments had to wait until the twentieth century. Notable voluntary rescues included saving Independence Hall in Philadelphia from demolition in 1816 and President George Washington's Mount Vernon in 1858. More important than the buildings themselves, however, was the organizational efforts of preservation advocates such as the Mount Vernon Ladies' Association of the Union chartered by Ann Pamela Cunningham in 1856 (Tyler 2000: 33). This private organization served as a model for the formation of other preservation associations that sought to save local landmark structures in the years that followed. In fact, many early groups were local history or patriotic associations, families, and government agencies that made efforts to save buildings not for their architectural import, but for patriotic reasons. While the early 1900s witnessed passage of several acts to establish the National Park Service and protect public lands and monuments, in 1935 Congress passed the Historic Sites Act enabling the Secretary of Interior to create preservation programs and hire unemployed architects and engineers to document historic properties. In 1949 the National Trust for Historic Preservation Act was signed to facilitate public participation in preservation, and in 1966 the National Historic Preservation Act (NHPA) became law and set forth the framework that guides preservation efforts today.

The NHPA established a series of institutions and procedures at the federal level for the protection of architectural landmarks and enabled the establishment of similar structures at the state level. Briefly, it created the Advisory Council on Historic Preservation comprising of twenty-three representatives from the private and public sector that advises the President and Congress. It also established the National Register of Historic Places, which is the nation's official list of worthy buildings, districts and sites, and the Section 106 review process to which federal agencies must submit if a project impacts an historic property listed on the National Register. The NHPA also established State Historic Preservation Offices (SHPO) to coordinate statewide inventories of properties, nominate properties to the National Register, and develop preservation and education plans. The state maintains its own register of landmarked properties which consists of all those successfully nominated from local jurisdictions.

Scholars argue that the 1966 act transformed the way in which Americans perceive preservation, which until that time had focused on established landmarks of great significance that could be turned into museums (Tyler 2000: 45). The NHPA broadened interest in historic preservation, encouraged the creation of local organizations and stimulated citizen participation in preservation activity. Moreover, it encouraged the preservation of a broader range of architectural structures, many of which had not previously been considered for "listing" on the National or State Registers. Yet, although both federal and state offices are important in registering buildings and supporting preservation activities, preservation professionals say that the local level is the most important for actually protecting structures and districts. Placing a landmark on the register does not guarantee that a building will escape modification or demolition. "Listing" only grants recognition of the property, but without the owner's assent to the listing protection is not ensured. Preservation professionals suggest protection can only be secured locally through passage of municipal historic preservation legislation, which is entirely voluntary. City councils are traditionally swayed by business people who often find that legal protections for historic architecture place too many expensive restrictions on the utilization of old buildings. Developers are well known for their strategies of purchasing devalued properties and deteriorated old buildings for demolition in order to build new structures.

Legislation developed at the local level can vary in scope and strength of enforcement. The differences are the result of local political forces that include elected officials, developers and business people, preservation advocates, and cultural groups. The legal structures put in place further affect the social dynamics in each community, privileging some interests over others, although acceptance of preservation tends to evolve over time. In general, cities pass legislation to protect architectural specimens of "historic" importance to the local community, labeling them landmarks. Some protections also include "districts," such as a "Main Street" commercial zone, or a residential area occupied by the city's founders. The legal authority for historic preservation legislation is grounded in zoning and land use regulations that are administered through municipal planning departments, sometimes also called economic development departments. Cities are usually required to set up a separate review committee or commission specially tasked with designating landmarks and districts, and approving modifications to historic properties. Often these review committees work in tandem with planning commissions or other municipal entities, and they may have a strong relationship with civic advocacy groups. In terms of what to preserve and how, municipal legislation is often modeled after federal and state guidelines in regards to criteria for designating properties according to their "historic significance," and following standards for retaining architectural "integrity."

The cosmology of significance and integrity

Two of the most important, and widely used, official conceptual constructs are significance and integrity, both of which, when implemented in the preservation process, rely on relatively subjective judgments. Significance is the primary requirement to be met to warrant protection; a building must have a significant historical association. As a minimal definition, significance refers to historical importance, but it is further defined in federal, state and local legislation. There are roughly four basic categories used to define a building's or site's significance in order to be designated worthy of landmark or district status: 1) association with historical events; 2) identification with important historical persons; 3) representation of distinctive characteristics of architectural style or craftsmanship, or association with notable designers or architects; and/or 4) having the possibility of yielding pre-historic or historic knowledge (Tyler 2000: 93–4). As these categories are further elaborated in state and local legislation, historical significance in terms of locality—the city or state—becomes more important. A house in a particular city might be deemed "significant" because a U.S. president once lived there, but it would be significant at local, state, and national levels. Another house could also be significant because a state congressman lived there, which would make it only important to the city and, perhaps, the state. In general, the more of the four criteria used to describe a property in a designation application, the more significance adheres to the project.

Historic preservation professionals describe certain other criteria in terms of their relative order of importance in designating a historic property as a landmark. First, the building usually must meet a minimum requirement of age, typically fifty years. In addition, its representation of the style and the extent of any alterations are considered, and lastly its association with historic criteria (Tyler 2000: 94). An exception to the age requirement applies to cases of relatively new, but unique examples of buildings that may be threatened with demolition. These factors tend to bias considerations toward old buildings built by elites, that represent "official" or unique architectural styles, are well maintained in their original condition, and associated with important events, people or designers. Vernacular architecture, the buildings built by and for common people are not often accorded the same consideration, and in the earliest days of the preservation movement were not even recognized as having potential significance. In recent years, many special vernacular forms have gained recognition as "folk" styles on their own or as collections to tell the stories of social movements or communities.

Judgments about whether an architectural structure is "significant" depend on subjective assessments made by local citizens, planning staff, historic preservation commissioners, and city councils that apply legally codified criteria to the proposal. Local debates about significance may arise in the designation process when actors ask if the property's history is significant enough, or if the significance is one that warrants protection. There seem to be few objectively measurable criteria provided to guide applicants or reviewers in this regard; rather, interpretations of standard criteria seem driven by local sentiments. Debates and conflicts about significance often reveal divisions within the city over "whose history" deserves preservation, or, alternatively, about who has the right to preserve (Tomlan 1998). Individual homeowners and developers tend to hold ideas contrary to those of preservation advocates regarding buildings worthy of protection. Quite often elite properties are preserved because they are presumed to be better built or are associated with official history, while properties meaningful to other groups are excluded (Green 1998). In fact, elites, and especially aspiring elites, frequently use preservation laws and techniques to exclude people and groups from occupying the same territories they claim. In some communities in the decades after the initial federal legislation was passed, the concept of significance was broadened and "democratized" so that the stories of many groups of citizens can also be preserved for later generations. While inclusionary policies have been adopted at federal and state levels, it is not always accepted at the local level. Many local municipalities with long-term relations with real estate developers remain hostile or indifferent to historic preservation and have resisted or ignored any attempts to adopt protective legislation or policies.

In practice, significance is closely related to another key preservation concept—integrity. Quite often debates revolve around the physical capacity of the property to "represent" the style or the history that is associated with it. Integrity refers most often to the material qualities of the building when it was first constructed—its "original" qualities, although it may also reference altered conditions associated with a later historical period deemed significant, or person or event argued to be worthy of protection. Integrity may be damaged if the building has been "modernized" in ways that destroyed its original features. Over time, most buildings undergo some changes, so one of the assessments important to considerations of significance has to do with whether the building or district is able to represent the history it is supposed to represent. Here, again, the assessments are subjective and are conditioned by considerations of proposed significance. If a building is badly damaged, but associated with a famous individual or event, it may still be accorded protection on the basis of the significance of the association. But the building would not be given such consideration if there are many other examples, or it were one of several buildings involved in the same historical event. On the other hand, some sites, such as the colonial capital of Williamsburg, are so significant that buildings that no longer exist have been "authentically reproduced" to flesh out the full picture of an historic place, and landmarked houses that have burned to the ground have often been rebuilt exactly as they were. Issues related to the restoration of integrity often raise questions about authenticity (Handler and Gable 1997).

Preservation professionals often feel compelled to preserve as much of the original material qualities of a building as possible. They refer to the Secretary of the Interior's Standards for Rehabilitation (1992), an official document of the Department of the Interior's National Park Service that provides guidance about how to conserve and rehabilitate an historic building. In fact, this document or portions of it are often cited or reproduced for inclusion in municipal legislation. This book is the preservationists' standard reference on techniques for the proper rehabilitation of a building's material qualities. It prescribes historic preservation's aesthetic ideals by making recommendations for resolving dilemmas with conserving or restoring original construction materials, replacing missing parts, making doors and windows usable again, and correcting structural problems. While the "Secretary's Standards" are especially critical in providing guidance to preserving landmarked structures appearing on state and federal registers, they are often used more loosely when dealing with local landmarks or properties in residential districts.

The concept of integrity is important to professionals and city officials in the designation process and when people engage in restoration practices, but may be less salient as a concept among homeowners and others not directly involved with preservation. When historic districts are designated, the integrity of the district must also be assessed through the survey of historic resources. Buildings that have not retained their integrity must be few enough in number so as not to threaten the integrity of the district. They are labeled "non-contributors" because they have been altered beyond recognition as an historic resource, or replaced with a newer structure. Once a district is designated, any proposed remodeling to a contributor property must be approved to make sure it does not destroy the integrity of the building or the historic integrity of the district. Acquiring official permission to remodel any historic structure, whether it is a landmarked building or a contributor to an historic district, requires review by the historic preservation board or committee. The reviewing body examines the proposed changes, advises the applicant, and grants approval, sometimes with conditions, in the form of a "certificate of appropriateness." That certificate is issued separately from any building permits and is often a prerequisite for consideration of proposals by the building department or planning commission. The state of California has adopted a California Historical Building Code (revised 2013), meant to guide and override applications of local building codes, which authorizes the retention of original features of the property.

State of California—CEQA Protections and Mills Act

While the philosophical issues surrounding "significance" and "integrity" are central to debates about preservation laws and limits on protective action, the state of California has also adopted laws authorizing additional protections for environmental and cultural resources. One law, the California Environmental Quality Act, or CEQA, was passed by the state in 1970 and provides governing bodies broad authority to protect cultural resources, such as historic architecture, from destruction. Although the CEQA law has played a minimal role in residential historic preservation activities on the local level,[1] the law was amended in 2006, and it revised the California Historical Resource Status Codes (CHRSC) that local governments could use to justify the protection of threatened historic resources. Based on a previous historic resource survey, a city could decide that a specific threatened property might be deemed significant enough for designation as a local landmark or contributor to a district and, therefore, could be protected even against the owner's will.

Another state law passed originally in 1972 was enabling legislation for cities to adopt ordinances aimed at protecting and restoring historic properties by private property owners. Under this legislation, popularly known as the Mills Act, the city enters into a contract with the owner of a landmarked or listed house, or a contributing property in a landmarked district, whereby the owner promises to restore the house in exchange for a reduction in property taxes. The contract has a minimum duration of ten years but can be extended. Usually, the homeowner makes a proposal documenting the property's architectural characteristics and a plan to restore "character-defining" features of the house with a schedule for completion of the tasks.

Local historic preservation histories

The City of Riverside (incorporated in 1883) was the second city in the state of California to adopt historic preservation legislation and it serves to illustrate over the long term how the movement developed in a community of traditional elites. As in many cities, such as Ontario, citizens became outraged when the city council authorized the demolition of the old Carnegie Library in 1964 to provide a plaza space in front of the new modern library. Many rallied to protest the council's plans to demolish other buildings that stood in the way of the city's redevelopment of its declining downtown area. To escape the wrath of activist residents, which included members of local elite families, the city council passed historic landmarks legislation in 1969, Title 20 of the Municipal Code, and established the Cultural Heritage Board as an independent reviewing body. In an attempt to distance historic preservation from city government, the Cultural Heritage Board was housed under the authority of the municipal museum, whose charter focused mostly on antiquarian functions of collecting historic documents and objects for display and educational purposes. None of the board members or museum's staff had any previous experience with building conservation or restoration. In 1977, the Cultural Heritage Board negotiated a grant from the state to begin surveying historic properties in the city stimulating, in turn, the creation of the Old Riverside Foundation (ORF), an advocacy group that argued in favor of land marking numerous civic buildings and protecting large Victorian homes associated with Riverside's late nineteenth-century citrus industry. According to some participants, there was little distinction between ORF members and the Cultural Heritage Board in the early years as membership in the two organizations overlapped. In 1984 the city's legislation was amended and updated, and in 1989 the city, recognizing the importance of preservation to economic revitalization, created planning positions to oversee activities related to the restoration of the Mission Inn, an important local institution, in the old civic center.

By the early 1990s, Riverside had made a fundamental shift in its perception of the proper place for preservation activities, and it moved the Cultural Heritage Board from the museum into the planning department. While planners had initially resisted overseeing preservation practices in the early years, by the 1980s, the Board had become a thorn in the city's side by blocking projects backed by development-oriented city councilmen. A former city staff member suggested that the Board had become more influential than the city council and wielded too much power. Another observer characterized the "old guard preservationists," whose missionary zeal and confrontational approach contributed to completing surveys and advocating for landmark status for many old buildings, but also as lacking knowledge of land use laws and due process requirements. As professionalization began to transform the preservation movement in the 1970s and 1980s, Riverside's Cultural Heritage Board members were increasingly challenged. They drew on history and archaeology faculty from the University of California, Riverside, as well as other at-large enthusiasts, for their knowledge of and expertise in preservation. While the faculty lent academic legitimacy to preservation efforts, the Board still lacked coordination and communication with the city in the realm of redevelopment projects.

Moving oversight of the Cultural Heritage Board into the planning department coincided with numerous changes in Riverside's preservation practices. One of the most significant was the incorporation of historic preservation goals, objectives and methods into various planning documents, including the city's General Plan. The constitution of the nine-member Board included individuals appointed by the mayor and city council, and required the inclusion of at least two professionals in architecture, history, archaeology, or urban planning. The state had also instituted the Certified Local Government (CLG) program, which focused on legally defensible preservation ordinances and required far more professional work to achieve preservation goals. The city hired more professionally trained staff and undertook revisions to its legislation and procedures in order to become a CLG program, which they achieved in 1995.

The city incorporated many changes to comply with other statutory requirements mandated by the state such as the CEQA, and changed requirements to allow for "administrative review" of minor projects. To allow for this, the city hired a Cultural Preservation Officer to conduct reviews, oversee the program and report to the Board. Although the city had conducted surveys of some 6,000 properties built before 1945 by 1980, by 2002 the database had increased to 9,000 surveyed properties. Citywide residential historic district design guidelines were initiated in 1995 and finalized in 2002. To accompany these guidelines, the city also adopted an education program with the Old Riverside Foundation, and engaged in outreach with owners of historic properties about the preservation program and owners' responsibilities.

By the end of the twentieth century, Riverside's city-wide historic preservation activities had become professionalized with highly trained city staff who held masters' degrees in public history or architecture, and whose programs kept pace with or took leadership roles on state and national levels. Civic organizations had also evolved over the same time. According to a former member of the planning staff, preservation in Riverside is likely to have originated in the early twentieth century with the local women's club, which was populated with the wives of landed gentry and other business owners seeking to protect the land and architectural heritage for their heirs. In 1927, the city hired architect–planner Charles Cheney to produce the city's first master plan, which introduced zoning to separate land uses. Cheney's wife, who was also the president of California's Federation of Women's Clubs, urged local women to become actively involved in preserving old buildings and natural landscapes. The buildings promoted as deserving of protection, however, tended to include only civic or large structures, and homes built by or belonging to the wealthiest families, the grand Victorians and architect-designed mansions. These elitist sentiments influenced the membership of the first activists who came together in the Old Riverside Foundation, and they encouraged the first survey volunteers, and architectural firms they hired, to focus on elite properties in particular neighborhoods and districts, rather than the ordinary vernacular or ethnically diverse structures associated with less affluent residents. According to one former planning staff member, the interests of the early advocates tended to focus on "pride of heritage," rather than preservation, *per se*.

During the 1920s, Riverside was divided into roughly twelve districts each separately housing white upper-class, middle-class, working-class, and nonwhite families; only two districts, Country Club and Oak Streets,[2] were inhabited by upper-middle class, or "bourgeois" residents (Tobey 1996: 87). Country Club included the hilltops of upscale Mount Rubidoux and slopes of the arroyo, where many socially and economically elite families resided, while Oak Streets housed many families from what one interviewee called the "merchant class" who were economically more mixed and socially integrated. These subtle social distinctions would come to play an important role later in the establishment of historic preservation districts, but the two residential areas were among the first to be surveyed. As a result of these surveys, a number of individual buildings were designated landmarks and, in the 1980s, the city proposed a number of historic "districts" to protect buildings that, on their own, lacked "significance" but as a group comprised worthy assemblages. The city carved out these areas naming them "Neighborhood Conservation Areas" (NCA), and then worked with property owners in the areas to formalize the designation of eight "historic districts" within or coincident with the NCAs.

Although some of the earliest historic districts were established in the city's center, two of the most important and, at the time, contentious historic districts to form were Oak Streets (1988) and Mt. Rubidoux (1990). In 1984, the city approached the eighty property owners in the Mt. Rubidoux NCA, where there were many architect-designed homes, about formally establishing an historic district. Cultural Heritage staff met with residents to explain the benefits and protections "districting" would provide to them and their houses. Many residents were favorably inclined, feeling it would "protect the neighborhood" and preserve the history of the area. One former staff member suggested theirs was an "elitist" position because residents really wanted to be able to restrict who was able to build on the remaining empty parcels in their neighborhood. Although a majority of residents wanted to exclude potential new residents, about a half dozen residents were adamantly opposed to the district, fearing that government was going to intrude or place restrictions on their private property rights. At the time, a federal regulation allowed owners of income property to take a 25 percent tax credit, which was not available to owner–occupiers (*Press-Enterprise*, May 15, 1984). This provision worried some homeowners who thought preservation would encourage more rental properties, while others were concerned the designation would attract sightseers (*Press-Enterprise*, June 21, 1984). Some opposed historic preservation because they thought it inhibited development, while others, especially retired military were reported to have said that they didn't fight in the military for this—that is, to lose their freedom. The six opponents of the historic district were able to convince the city council to block the move. Several years later, after a former State Preservation Officer moved into the district, sentiments changed and in 1990 the district was designated.

The situation with Oak Streets, where interviews for this study were concentrated, is more complex. An NCA was established in 1981 with the survey, the Cultural Heritage staff had advised residents about the possibilities of establishing one or more historic districts. The whole of the Oak Streets NCA included some 1,224 properties of which 1,081 are considered "contributors"—a very large area mostly built between 1910 and 1940. By 1984, between surveys and city efforts to create other residential districts, many residents had heard of historic preservation. A Cultural Heritage staff member received a phone call from a resident of Rosewood Place in the NCA asking why Oak Streets had not been made an historic district. He responded by describing the conflicts the city had encountered with Mt. Rubidoux residents. Her response indicated that all the residents of her street were by this time ready, which quickly resulted in establishing the first historic district called Rosewood Place in 1986. Working with the remainder of the NCA proved to be more of a challenge for Cultural Heritage staff.

A group of residents from the north end of the Oak Streets NCA clearly wanted their neighborhood to become an historic district. They claimed a strong sense of identity and belonging, and promoted themselves as having an intense social commitment to their neighbors. At the June 1984 meeting of the Cultural Heritage Board two residents presented a petition signed by 100 of their neighbors to become an historic district. In a 1986 newspaper article, Oak Streets was described as a traditional, family-oriented neighborhood where people move in and stay. This article also claimed that it had a unique sense of place recalling an earlier simpler time when neighbors "borrowed cups of sugar from each other," helped each other, and sat out on their front porches (*Press-Enterprise*, September 10, 1986). Interviews summarized in the newspaper article revealed that residents not only knew each other, but they also knew the histories of their homes. Some recent arrivals from the newer community of Irvine in Orange County contrasted their new life in the Oak Streets with their previous residence where they never saw the mailman and no one took a walk in the evening. The article pointed out that over 90 percent of the residents were white.

The petitioners for the historic district, however, did not want to include all houses in the NCA, but only 195 properties at the north end. The chair of the Cultural Heritage Board at the time, an architect who also lived in the Oak Streets NCA, thought it would be a travesty to break up the entire NCA. He said that he had undertaken an informal survey on several streets near his home, and discovered that his neighbors were not all that interested in preservation protections. He suggested that while people had maintained or personalized their houses through remodeling, other properties had deteriorated from lack of care or from being rentals. According to him and others, the houses in the northern section included some very impressive houses, although not all of them were grand like the houses in Mt. Rubidoux. In fact, outside the northern edge of the NCA the houses were of more modest vernacular styles. But he also noted that no one had asked him to sign the petition for the district, and he likened the dispute about dividing up the district to the cliquish conflict that had occurred in Mr. Rubidoux. The north-end petitioners sought the designation as an historic district to set themselves apart from the rest of the NCA, to acquire an elite standing akin to what the city had offered to residents of Mt. Rubidoux. It was an exclusionary distinction they continue to assert.

The official designation criteria for the NCAs and Historic Districts worked their way into the arguments advanced by the petitioners. Nominations for designations are made by the city council, Cultural Heritage Board, or petition by a property owner(s), but must be approved by the city council. As outlined in Title 20 of the Riverside Municipal Code (Chapters 20.05–20.45), the designation of individual buildings and neighborhoods was at the time two-tiered, reflecting the early, perhaps elitist inclinations of historic preservationists who did not recognize the value of vernacular architecture. For individual buildings, a "Cultural Heritage Landmark" is a "cultural resource of the highest order of importance," while a "Structure of Merit" is important, "but at a lesser level of significance than a Cultural Heritage Landmark" (Riverside, Citywide Residential Design Guidelines 2003: 14). Likewise, an Historic District is defined as a geographic area with "a significant concentration of cultural resources that represent themes important in local history," while a NCA is defined as similar, but with "structures/resources of somewhat lesser significance" or of "lesser concentration" than a district (Riverside, Citywide Residential Design Guidelines 2003: 14). These distinctions continued to be used in the city until 2006 when the Municipal Code was revised and the two-tiered definition of historic district and NCA was eliminated. Thus, the distinctions initially created in the legal language of the municipal code might also have been appropriated and employed to mirror social divisions within the larger community. While elites in charge of writing that original language were probably focused more on material qualities rather than social ones, it is no surprise that social exclusion is still expressed through historic preservation designations twenty-five years later.

In 1988, the Oak Streets Historic District, comprised of seven and one-half blocks of houses was carved out of the NCA and established. The significance of the district emphasizes "the cohesive representation of typical southern California residential architectural styles, circa 1916–40" (Riverside, Oak Streets Historic District). Technically, the significance of the district is virtually identical to the description of the NCA. According to Riverside's historic preservation officer, the distinction between the two types of designations does not make much difference in how properties are treated. Any proposal to remodel requires that homeowners apply for certificates of appropriateness regardless of the type of district they are in. Interviews with some residents of the district revealed the notion of "lesser significance" to characterize houses in the NCA as having less economic and social value, less prestige, but not necessarily lesser historic value. That is, the market value of those houses is seen as lower, and the types of people who live in them are perceived to lack the higher-class status. These characterizations also appear in the organization of preservation groups, which will be discussed in the next chapter.

Although Riverside curtailed the establishment of any new NCA designations in 2006, the four earlier designations continue to be recognized, which presents continuing problems for the city. Many Riverside residents, including some that live in an NCA, do not or did not realize that there is a distinction between the NCA and Historic District designation. More important, however, is the longer-term problem of what to do with the NCA properties—do they join the historic district, which is clearly being rejected by the current homeowners, or can some other arrangement be found? City officials would like to see the entire NCA included into a single formally desig-nated Oak Streets Historic District, but there is strong resistance from some of the historic district residents. On the other hand, it could become its own historic district, adding one more to the current total of thirteen. At the moment, this conundrum has not been resolved, but it serves to illustrate the exact ways in which preservation ideas such as "significance" and "integrity" become embedded in the exclusionary discourse of place and social relations.

Experiences with historic preservation legislation in other cities

Some fourteen miles west of Riverside is the City of Ontario, which became a Certified Local Government (CLG) in 2003 with an updated version of their historic preser-vation legislation originally passed around 1993. Ontario's preservation program falls within the orbit of Riverside having been influenced by residents and professionals who were educated in UC Riverside's public history master's program. Surveys of Ontario's historic resources had been conducted in 1983 and produced a list of some 3,000 properties, both large public buildings and homes, with historical significance. The survey also identified neighborhoods with enough consistency to establish historic districts like Riverside and buildings that could be landmarked, but Ontario was not eager to adopt protective historic preservation legislation. In the early 1990s a citizen's commission was appointed to study and propose legislation, and a proposal that met state standards for local protections was put forth. According to one participant, few long-term residents took much of an interest. The city council, which was very pro-development, felt the restrictions placed on property owners in the proposed legislation were too strict and would inhibit development. The committee had proposed a body to review nominations and grant certificates of appropriateness that would be constituted independently of the city's government apparatus, much like Riverside's Cultural Heritage Board. As a result of the "extreme" proposal, the commission was disbanded, and by 1993 the city had adopted a weaker version of the proposed legislation. The new ordinance constituted the reviewing Historic Preservation Commission as a subcommittee of the Planning Commission, a practice roundly condemned by preservation professionals.

Operationally, the legislation worked in ways similar to Riverside's procedures with the city outlining potential residential districts based on survey data, and then meeting with residents to formally designate them. There are currently six districts, but another five are proposed, and yet another five have been identified as potential districts. The current law provides for "Architectural Conservation Areas" similar to Riverside's NCAs, but none have been identified. Unlike Riverside's experience, the kinds of houses that appeared in the original 1983 survey included many "vernacular" examples of modest homes as well as large Victorians and Craftsman houses. In 2003 when the original legislation was updated, a new survey was conducted in order to re-establish the historic resources originally enumerated, and to "de-list" any properties that had lost their "integrity" due to unchecked remodeling processes. Planners found that because the city had failed to enact or enforce protections in the twenty years between the original survey and the resurvey, almost half of the properties had suffered some kind of alterations that required them to be deleted from the original list.

One former Ontario city planner observed that most of the houses that were "de-listed" were located "below the railroad tracks," in areas associated with a large lower-income Latino population. Homeowners in these areas have routinely applied stucco to their wood-sided bungalows and enclosed front yards with wrought iron gates and fences. From a preservationist perspective, the houses have lost their integrity and not enough remaining properties considered "contributors" can be found to constitute districts—nor is it clear that any of the residents would want that. All Ontario's designated residential districts are located in the northern part of the city, where there are many more large mansions, Victorians, and Craftsman houses inhabited by many old guard families or wealthy business people and city employees.

The fact that their residential districts were established while much of the downtown area was being demolished to make way for new construction, speaks to the pro-development sentiments of the city council and the economic interests of their supporters who live in protected districts. Still, without much active interest or advocacy from Ontario residents, historic preservation activities have been largely directed by the city's professional planning staff, who have elevated the city's reputation among regional preservation professionals. Slick brochures and awards programs, database construction, and well-organized websites are notable accomplishments, but Ontario's residents generally appear apathetic.

Ontario's experience with significance and integrity in historic preservation has resulted in a very mixed bag of historic building stock. Although the finest old houses

Figure 16 Residential district marker, Ontario. Credit: Denise Lawrence-Zúñiga.

singly and in residential districts have been retained, and some historic civic and large commercial buildings have been preserved, the bulk of the downtown has been under threat and many buildings have not survived. One planner argued that the unrein-forced brick buildings built at the turn of the century in the city's commercial core can not be salvaged. She suggested the cost of making them seismically safe is prohibitive to developers. Still, the cities of Riverside and Pasadena have been able to ensure the preservation of many early brick buildings in their downtown areas by insisting that developers undertake those improvements. These contrasts suggest that Ontario's city council and their planning department are not inclined to require investors and developers to spend the money necessary to protect the historic fabric. As one activist pointed out, the people who are most likely to get elected to city council positions are those that receive the most money from developers, the developers who actively promote the idea that newer is better. In this regard, the Planning Commission's Historic Preservation Committee seems almost eager to approve projects that conform to the development ideal promoted by the city council and Ontario's elite families.

Further west to Monrovia and Pasadena

In both Riverside and Ontario, historic preservation activity seems to involve relatively few activist residents, but the cities themselves have developed highly professionalized preservation services and programs. Monrovia, located just twenty-six miles west of Ontario, and Pasadena, some thirty miles west, are also cities with professionalized preservation programs, but with more energetic preservation activism and a different civic dynamic. Both cities adopted historic preservation legislation that allows them to designate landmark buildings and historic districts, but without referencing the greater and lesser architectural qualities found in Riverside's code. Monrovia, however, has only one residential district established in 2009, while Pasadena has at least twenty-seven residential districts, the earliest dating from 1989. Monrovia adopted its legislation in 1995 as part of their municipal code's zoning regulations, but Pasadena had already passed historic preservation legislation as early as 1969, the same year as Riverside. Pasadena residents have used residential districting as an exclusionary practice to curb development practices, while Monrovia has devised other tactics including survey listings, CEQA regulations and negotiation to gain some control.

Monrovia's historic preservation legislation embodies the same concepts of significance and integrity as other programs. It also includes strong statements about requiring the consent of the property owner for the designation of a landmark. In the case of districts, it requires 50 percent plus one of the property owners to agree to form a district. As in other cities, the Historic Preservation Commission's role is described in multiple sections as "advisory," and all decisions are subject to appeal to the city council. As if fearing government imposition, however, Monrovia's preservation community emphasizes that these requirements make their program "voluntary," even though most other cities have identical requirements. According to some members of the original Historic Ordinance Committee formed in the 1980s, the city council issued a very strong warning that if the proposed legislation required mandatory participation, so as to compel unwilling property owners and especially developers to preserve buildings, they would not approve it. Committee members pointed to discussions at the time about the neighboring city of Sierra Madre, which had strong requirements for developers and owners that provoked residents to protest. Eventually that city had to modify some requirements to placate vocal citizens. As a result, Monrovia relies on property owners to apply for the landmark designation of their houses, a process that is not always successful. Moreover, there is some reluctance to designate districts even when, technically, only 50 percent, plus one of the owners are required to consent. In addition, as the city planner explained, because houses were built as infill over many decades there are few neighborhoods that show enough visual or stylistic consistency to warrant designation. One neighborhood may have houses from every decade from 1910 to 1970.

Although Monrovia's preservation ordinance states that designating districts requires only a majority of the residents to agree, it is interpreted as potentially coercive, not voluntary, and, so far, achieving consensus seems to be preferred. Residents of the first district began discussions for forming a district in the 1980s, but are said to have waited years before everyone was on board before consenting to a district designation in 2009. The Wildrose Historic District met the "significance" criteria in four of seven categories, and demonstrated high levels of integrity with nine of fifteen houses having already been designated individual landmarks; only three houses were declared "non-contributors" in the Wildrose district. Because the original

1985 survey of historic resources did not show any other potential districts or cohesive areas, it is unclear whether the city, in promoting the idea of voluntary participation, wanted the residents to achieve consensus first, or the residents of Wildrose just wanted it themselves, or perhaps both.

The city has also made proposals for additional districts, and has met with some of the property owners. The initial boundaries cautiously drawn for one district brought complaints by neighbors who were left out, which led to expanding those boundaries to include more homes. Deliberations about designating or remodeling historic properties, however, occur at the regular meetings of Monrovia's Historic Preservation Commission. The Commission consists of seven individuals chosen by the mayor and city council and includes knowledgeable professionals (from the same categories as Riverside specifies) such as architects, realtors and preservation activists. Historically there has been some overlap between the civic preservation organization and commission membership, and the civic group often has a highly vocal presence at the commission meetings when there is a particularly controversial issue to discuss. According to local residents, meetings are generally pretty calm and uneventful, however, some issues have become contentious.

One of the more contentious issues arose, in 2010, over a Victorian-era property, an 1896 folk Victorian farmhouse that stood on a double lot. Two Monrovian developers purchased the property with the intention of demolishing the structure, subdividing the lot and building two new houses. Because the old house had been listed in a prior survey, its demolition had to be reviewed by the commission. The commission's powers do not allow it to completely block demolition, but only delay it for up to 120 days. The delay is meant to be a "cooling off" period during which the interested parties can try to reach some agreement to amend the project proposal. After multiple meetings between the commission, the civic advocacy group and the developers, the developers agreed to retain the original structure of the house and add onto it, while still subdividing the double lot and building an additional new house. In the construction process, however, the developer was only able to save little more than the façade, finding much of the structure termite-eaten or destroyed by dry rot. In the end, the façade and much of the original house had to be reconstructed, but not without provoking the ire of the commission and local activists.

In discussing the final but unhappy outcome, preservationists and commission members acknowledged that their legislation allowed them no powers to protect a resource. The house was not considered "significant" enough in terms of associations with individuals or events, or in terms of its architectural style. The "Folk Victorian Farmhouse" designation was a kind of consolation prize to preservationists and, even though there are few of these still standing, it was not enough of a reason to grant it "significant" or landmark status. Monrovia's local historian complains that this is partly due to a "great house (man) theory of protection," which he traces to the early days of preservation in Monrovia. He says that conducting the first survey focused on identifying and saving particular examples of excellent architecture while overlooking smaller and less important houses that constitute the fabric of the community. A similar complaint is also heard from the many homeowners who have applied for landmark status for their houses in order to get a reduction in property taxes through the Mills Act, and have been turned down because the history of their house is not deemed "significant" enough. Indeed, many of these owners suggest that the Commission is elitist, and that the criteria they use sets too high a bar.

These and other complaints have generated some discussion in the community and on the Commission about legislation, especially with regard to vernacular houses that fall through the cracks, many of them sited on the double lots so attractive to developers. This incident also raises questions about significance; what constitutes a truly historic house? Without districts to provide generic protection to all the contributors, an evaluation of the individual house is required. Here, the planning department uses the listing on the survey of historic resources to determine the house's potential for landmark designation with local significance. The department combines that with a "Status Code" provided by the State of California using CEQA law. The categories refer to the eligibility for landmark status, although some categories also suggest that a property lacking local significance might deserve special intervention by the planning department. The farmhouse was one of these properties. Once the failed remodel and addition was complete, the farmhouse structure had lost whatever integrity it previously had. Some residents were very upset because they felt helpless to intervene under current law. While a district designation could remedy the situation and provide a modicum of protection, according to the city planner, Monrovians tend to see districts as difficult to designate because of the eclectic content of neighborhoods. In addition, they are reluctant to use the 50 percent plus one consent requirement, because it seems potentially coercive rather than voluntary. In the meantime, additional old houses are being issued demolition permits, while the owners of still others are engaging in restoration that does not retain the original historic integrity.

In the city of Pasadena, the Historic Preservation Commission is similarly constituted of nine members selected by the mayor and city council, but with additional representatives from selected historic districts. The planning department has seven staff members devoted to design and historic preservation, with specialist staff members providing services to the commission. Pasadena's adoption of historic preservation legislation in 1969 allowed the city to begin immediately protecting its architectural resources, including in the 1970s many of its civic buildings and large mansion-sized houses designed by architects such as Greene & Greene and Frank Lloyd Wright. In addition, bungalow courts or courtyard housing, which were threatened by demolition and apartment construction because of their larger lots, were also studied and designated. It was not until the 1980s, however, that the impetus for designating residential districts of vernacular housing began. The first neighborhood district was established in 1989 in the older northwest neighborhood of Cottage Heights.[3] Houses there had suffered from deterioration, gang infestation, and replacement with large apartment buildings. By 2009, twenty-seven self-organized landmarked historic districts, commercial and residential, had been established in the city.

Although the city had surveyed selected areas throughout the city, and worked to designate landmark buildings and some districts, much of the city's oldest residential building stock remained unsurveyed. Pasadena's neighborhoods did not suffer from the same problem of period inconsistency as did Monrovia's with houses built over five or more decades, nor did residents or the city council believe it coercive to require 50 percent plus one of the residents of a proposed district to be in favor of forming a district. In fact, historically, Pasadena's city council has been favorably inclined to historic preservation protections. So much so that many preservation oriented activists have discovered that creating an "historic district" can be used in an older neighborhood to not only protect specific properties from being irrevocably altered through demolition and construction, or remodeling, but also to protect the entire neighborhood and its historic character.

As an example, in 2005 the Holliston Avenue Historic District, consisting of twenty houses on both sides of a street, was designated. The move to establish the district came as a reaction of local resident-activists to an increasingly common threat from real estate developers who were targeting some of Pasadena's older unprotected neighborhoods. Holliston Avenue is the site of a 1936–7 bungalow court of not very distinguished qualities, but it is surrounded by older bungalows and homes built as early as 1907. Early in 2004, the new owners of the court, a Chinese couple from nearby San Marino, and their architects, filed an application to demolish the bungalow courtyard in its entirety to construct a condominium complex with eight townhouses; there already existed some apartment buildings on the street. Seeking protection for the courtyard, one of the former residents and her neighbor consulted with preservation activists, and gathered signatures in support of an application they wrote for the neighborhood to become a Landmark District. They submitted their application in July 2004 and by September the developers had revised their proposal to demolish the entire complex and proposed instead to demolish all the trees and one rear cottage to build a two-story four-unit addition with parking. In September, the two advocates nominated the courtyard complex for individual Historic Landmark status for additional protection, and submitted the application to the planning department.

Designation of the Landmark District proceeded without any hindrances, although it took until July 2005 for the proposal to wend its way through the Historic Preservation and Planning Commissions before arriving at the city council for final approval. The fate of the proposal to make the courtyard itself a designated Historic Landmark ran into more trouble. Planning staff reviewed the application, conducted their own research, and concluded that the property did not meet the significance criteria for designation as a landmark. One of their observations was that the bungalows were built well after the 1931 cutoff for courtyard "significance." The applicants appealed to the Historic Preservation Commission, which then voted unanimously in November 2004 to approve an Historic Landmark designation for the courtyard. The Commission said that the property was significant because of "its simplicity and lack of ornamental details and its courtyard plan" (Pasadena, Historic Preservation Commission, Agenda Report, April 25, 2005). According to one of the applicants, the Historic Preservation Commission agreed with them about the courtyard's unique humble qualities as housing for ordinary working-class Pasadena residents, and that its simplicity spoke to the economic hard times during the 1930s. In June 2005 the Planning Commission also approved the landmark designation for the courtyard property. The developer did not appeal these decisions, and the proposal went to the city council for a final determination.

The planning department then hired an outside professional consulting firm to conduct their own independent study, and they concurred with the planning staff that the courtyard houses did not exhibit significant characteristics consistent with criteria for designation as a landmark property. In their opinion, the problem with the courtyard was that it was built near the end of the depression, after the period of courtyard construction, and did not "possess high artistic values" because it was simply built and lacked any special design features (Pasadena, Historic Preservation Commission, Agenda Report, Attachment "A," April 25, 2005). While the courtyard landscape was notable, and featured a number of mature specimen trees, that was not enough to warrant protection. Now the project appeared before the city council for a decision to resolve the conflict between the planning department recommendations and Historic Preservation Commission decisions. The developer submitted substantial documentation in the form of letters of support and two additional independent reports on the property conducted by other historic preservation professionals. On July 18, the city council reviewed the District Landmark application, which it approved, but denied the Individual Landmark application.

Although the two activists were unable to secure landmark status for the bungalow courtyard property, they were successful in securing District Landmark status for twenty properties on Holliston Avenue, which included the courtyard. District Landmark protections are not as strict as those for an individual property, but they certainly enabled a good portion of the property to be preserved. Other neighborhoods throughout the city have employed similar tactics in order to resist development threats. In one upper middle-class neighborhood of older homes, developers acquired two newer postwar houses with plans to demolish and add outsized second floors as speculative projects. The neighbors successfully organized and proposed an historic landmark district to block the developers' efforts. "Districting" has become a potent strategy, so that real estate developers who have enjoyed success in neighboring communities tend to avoid using the same strategies in Pasadena.

These cases also demonstrate that motivated residents learn very quickly how to understand legal requirements and use the preservation language to advocate for their positions. While they may lack the depth of knowledge and experience professional preservationists, historians and architects employ, many can pick up enough practical knowledge to become credible activists. Moreover, they become impassioned advocates for the issue because it holds personal meaning for them. There is frequent mention in professional reports that lay people often have misunderstandings about historical "significance." In a memorandum to the Pasadena Planning Department dated April 2005, professional consultant Leslie Heumann wrote, "[T]here is a [*sic*] often a tendency among advocates for historic preservation to confuse 'history' with 'historical significance.' That is, all properties have a history, but simply documenting that history does not provide evidence of significance" (Pasadena, Planning Report, April 18, 2005). The memorandum goes on to explain that the determination of significance is outlined in the *National Register Bulletin 15: How to Apply the National Register Criteria for Evaluation* (1992), also known as the "bible" to preservation experts. It could be argued that local significance is far different from national significance, and that local communities may consider an otherwise undeserving property to be highly significant and meaningful to them.

City of Alhambra—preservation without legislation

Of the five cities considered in this study, only one, Alhambra, has never adopted historic preservation legislation, but it is not because local residents have not tried. Rather, preservation advocates in the city have pressed for the adoption of legislative protections since 2003, but the city council which, like the city of Ontario, has had historically strong connections to real estate developers has resisted. Instead, the city council, in 2008, finally agreed to pursue residential design guidelines as a partial measure to protect historic residential resources. Alhambra's oversight of planning and building activities is housed in the Development Services department and is served by a Planning Commission that administers zoning regulations, and Design Review Board (DRB) which is headed by a staff architect. The staff architect and planning staff provide guidance to designers and property owners, but official reviews by the DRB are required for every new or remodeled architectural project. The preparation of the Residential Design Guidelines was originally contracted out to a private planning firm, but once they were officially adopted in 2009, they have been administered by the DRB. The Board is comprised of six members nominated by the six members of the city council. It includes architects, landscape architects, developers and interested citizens. It meets once or twice a month, depending on the workload, and, like historic Preservation Commissions in other cities, is open to the public.

Unlike historic preservation legislation, residential design guidelines do not require property owners to conform to the specific design features illustrated in the guidelines or recommended by staff or DRB. That is, the guidelines are "voluntary" and the way recommendations are incorporated into designs can be negotiated. This means that the guidelines are a lot less stringent than preservation laws and they permit designers room for interpretation. Nevertheless, the guidelines themselves are structured along the same thematic lines as historic preservation legislation in that they emphasize the designation of historic areas and provide detailed suggestions about retaining stylistic integrity in design. The guidelines identify and name residential neighborhoods by their historic significance and common architectural features, but lack the same legal protections found in other cities with preservation laws for all the houses. Moreover, these neighborhoods were "constructed" by the planning consultants and were not based on the consensus among the residents. This lack of buy-in by residents presents continuing problems for preservation advocates, property owners and the DRB.

Alhambra's partial solution to the historic preservation issue can be attributed to the persistent efforts of the city's preservation group, the Alhambra Preservation Group (APG). This citywide organization was originally established in 2003 and began advocating for protective legislation by attending and participating in DRB meetings. The consistent and active participation of several well-informed APG members who spoke in meetings to support recommendations of DRB members, or added additional historical research information, actually won their trust and that of the staff architect.

APG's advocacy was critical to raising awareness in Alhambra and will be discussed in some detail in the following chapters.

Conclusion

Historic preservation cosmology is officially constructed in legislation and the scholarly and popular work of nationally recognized professionals and organizations, just as homeowners and civic organizations use it "unofficially" to justify remodeling practices and engage in advocacy at the local level. Concepts of significance and integrity, for example, are the core ideas for defining legal requirements and guidelines, but actual implementation is subject to interpretations according to localized histories and traditions, as well as sentiments and judgments of appropriateness. The local historic preservation cosmology is socially constructed and reproduced within an active context of political, economic and social realities. Strict adherence to codified historic preservation concepts that are written into local legislation and regulations can vary from city to city, and over time, as new pressures and threats arise. Some variations appear in the structuring of municipal offices and commissions that administer historic preservation laws while others are expressed in the public arena in which community preservation groups tend to play a significant role.

The construction and reproduction of the historic preservation cosmology among local homeowners and professionals ensures their participation in a community of believers in the value of preserving for posterity architectural resources representative of a city's past for posterity. This community extends far beyond the locality and unites people who share knowledge and interests. It is also an aesthetic community bound by the tacit agreement that the representation of historic significance is to be expressed in stylistic features of built form. The cosmology constructs a kind of aesthetic normativity that guides owners, builders and officials whose actions strive to implement the material ideals, thus making historic preservation a moral project for the city, for its identity and its historic meaning. In referencing the cosmology in everyday activities, even if imperfectly, homeowners become empowered as they link their efforts to a social and aesthetic movement greater than themselves and even their own city. The next two chapters explore local organizations and their advocacy activities in four cities in the study and focus on how residents use preservation laws inclusively to construct the boundaries around their own communities and exclusively to separate themselves from others.

Local Level Preservation and Exclusion: Traditional Elites

The likelihood of a city adopting historic preservation legislation, or even having an interest in its architectural past, is more often than not the result of local residents' advocacy efforts. During the late nineteenth and early twentieth centuries, concerned citizens participated in historical societies or museums, organizing them to act as repositories for documents, archives, photographs, and memorabilia associated with the founding of the city and its early leaders. Sometimes these collections were housed in the city's main library and were curated by especially devoted librarians. The wives of local business and political leaders often formed women's clubs and were instrumental in promoting the collection of material culture. As in the case of Riverside, these resources formed the core holdings of museums and historical societies, but there was little civic interest in a city's architectural heritage. Rather, the focus on historic architectural resources often came as a consequence of a tragic demolition that triggered concern. The disastrous effects of urban renewal and modernization on many early twentieth-century buildings like the Carnegie Libraries (1883–1929) tended to make them particularly good candidates for stimulating local preservation interests. Often viewed as eyesores after decades of service, they and other prewar buildings seemed ripe to city leaders for replacement, but once they were demolished citizens longed for their beloved monuments that recalled an earlier age. The evocation of a sense of loss, of nostalgia, coupled with fears of potential future losses are what commonly motivate citizen advocacy for historic preservation protections.

In southern California's suburban cities, these sentiments most often began to appear and strengthen during the postwar period when urban renewal projects targeted older neighborhoods for modernization. In fact, citizen passions were most fervently directed at real estate developers whose proposals to build new commercial and residential projects coincided with the interests of local politicians. In many suburban cities established in the early twentieth century, vacant land for construction was difficult to find in the urban core. Thus, already developed areas with deteriorating building stock looked particularly attractive to investors, especially if those places were also associated with other economic and social problems. To this day, local politicians imagine a solution to socioeconomic problems through redevelopment, or more euphemistically "revitalization," that requires demolition and rebuilding. Prewar cities are particularly challenged to retain or recapture their economic vitality as new developments on the periphery pull economic resources and activity away from the center. Some redevelopment strategies emphasize building new urban cores in the same styles as the new suburbs, but more recently some older cities have discovered that "branding" their identities by restoring their historic urban cores can help distinguish them from newer urban centers.

Developers are not the only threats to retaining historic architectural resources as private buyers of affordable but deteriorated buildings also envision an alternative suburban landscape. Recently arrived affluent immigrant groups may also see in older buildings past their prime and ready for rehabilitation opportunities for expressing a new aesthetic character and establishing a new community with a new identity. Preservation advocates, however, argue that expanding and adding decorative architectural features that destroy the original features does just as much damage to historical architecture as demolition itself. Still the economic renewal that private investment brings, regardless of the architectural aesthetic, is highly prized by local politicians and business elites.

Some of the most vocal opponents to demolition, and developers' and private owners' design practices usually include residents who embrace historic preservation values. As we have seen in previous chapters, individual homeowners personally construct their own preservationist identities and lifestyles pursuing remodeling practices at home. To varying degrees, from purist to practical, homeowners interpret aspects of the preservation cosmology to justify the restoration of the original material conditions of their houses. But the history homeowners preserve is experiential, rather than "official," and the remodeling practices that give their everyday life new meaning tend to be more personally transformative than factually accurate. Even so, homeowners do not act in isolation from other influences but engage home remodeling in a social context that includes learning from neighbors and the larger community. In learning restoration techniques or the value of architecture history, homeowners seek involvement with others who share their appreciation and want to promote their shared values to others. Joining existing historical societies or establishing a new organization provides a means to find like-minded residents in the same neighborhood or in the city at large. These voluntary associations engage in education and promotion through home tours and special events, they provide lectures and workshops, connect homeowners with restoration resources and each other, and engage in advocacy at city hall.

It is often the case that the same homeowners who have enthusiastically embraced preservation practices in their private lives also express intense preservation sentiments in the public sphere. The most active advocates press municipal governments and other large civic institutions to promote the preservation of a city's architectural heritage. Their participation in local civic affairs may be seen as a simple extension of the cosmology that motivates and explains their private acts of restoration into the public sphere, but it is more complex. The extra-domestic focus of the cosmology aids in the construction of a community imaginary, an idealized, romanticized historic landscape that can be reproduced or resurrected through restoring architecture and passing preservation laws. The historic landscape of any one city is constructed through the collective efforts of preservationist homeowners and other historically minded citizens who want to see the local architectural heritage preserved and honored. This historic imaginary, however, is largely "invented" and contests other idealized landscapes, modernized or immigrant enclave, promoted by developers and city councils committed to remaking their cities to attract more businesses and tax revenues. Even in local neighborhoods that have obtained protected district status, residents still find developers threatening to buy and rebuild, or neighbors with remodeling projects that challenge the residential imaginary. These potential and actual contestations motivate preservation homeowners to become involved in neighborhood and community affairs through local organizations.

The preservationist community is a largely symbolic one that brings together individuals of common interests and sentiments while contrasting them against others with different ideas and opinions (Cohen 1985: 12). The preservationist community is place-based; it interprets and seeks to protect the qualities of a particular space for its historic and aesthetic values. The boundaries of this community are "material," but not in the sense that they refer to a geography that literally contains or includes its members. Rather, it is based in the symbolic representations of history found in the built environment of a particular place, or any place, that people use to mark themselves as adherents of the preservationist ideology and distinguish themselves from others. Use of these symbolic referents, however, requires knowledge and discerning judgment, or taste, to effectively manipulate the symbols in a convincing way. Taste is learned at home and is part of the cultural habitus associated with class identity (Bourdieu 1984: 34). Discerning aesthetic judgment learned through social and material contexts and "naturalized" or taken-for-granted, binds together people who share in the unquestioned sensual pleasure of particular things, buildings and landscapes (Duncan and Duncan 2001: 392). Even so, it is clear that many preservation homeowners learn about architectural features signifying the authenticity of historic buildings and styles only when they begin remodeling their own houses. That is, their "discerning judgment" is acquired in the process of restoration. Thus, the taste that homeowners are expected to perform to signify social standing may be acquired much later than childhood, even if enabled by an early educational awareness typical of middle-class enculturation.

It would be a stretch to call the acquisition of an historical aesthetic sensibility a necessary indicator of class membership, since there are many members of the upper-middle and middle classes who express other aesthetic preferences, or who entirely reject preservationist values. A more exact characterization suggests that members of the middle class generally ascribe aesthetic value to the material qualities of domestic architecture and interiors, but vary in terms of the styles they prefer. While some homeowners prefer mid-century modernist materialities, others prefer Craftsman or Victorian style houses. This suggests that those sharing aesthetic preferences for a particular style or historical period might constitute a "status group" implying common interests and lifestyles. Status groups, according to Weber (1948), "attempt to monopolize the supply of honour, recognition and respect" (as quoted in Redfern 2003: 2359). In contrast to other homeowners, developers and local politicians, preservation advocates differentiate themselves philosophically to claim recognition and respect, and assert the moral high ground by suggesting that their aesthetic concerns are mostly focused on protecting architectural heritage and the history of the city. These notions accentuating a sense of honor and moral superiority reinforce a kind of exclusivity that has been traditionally associated with elites' historic preservation causes, even if these sentiments of exclusivity are "naturalized" by adherents rather than consciously promoted.

The construction of the symbolic community around shared aesthetic preferences necessarily excludes others; indeed the very construction of preservationist tastes is often based on the collective and conscious rejection of selected mass-produced, industrialized domestic commodities found in contemporary society. It is the conceptual contrast between preferred and rejected tastes in constructing the symbolic community that forms the basis for social exclusion. As James and Nancy Duncan argue, "exclusion acquires an aura of scarcity and thus becomes a form of cultural capital ... An exclusive neighborhood thus is a positional good, one that is highly sought after" (2001: 390). While preservation homeowners may seek exclusion by establishing geographically distinct neighborhoods populated with historically significant styles, they are much less conscious of the effect of these practices on people that do not subscribe to their aesthetic values who are physically excluded from the neighborhood. The rejection of other sets of aesthetic values, however, goes much deeper than simple physical segregation and zoning of differences because alternative aesthetics expressed in close geographic proximity can threaten the integrity of the whole historical landscape conceit preservationists attempt to revive or reconstitute.

All neighborhood "districts" accorded historic preservation protection are exclusionary by definition, largely because legislative regulations are written to control and prevent certain homeowner remodeling and design behaviors. But there are different kinds of exclusionary tactics at work among the residents. For elites who have maintained continuous residence in one place over decades, preservation may be more about keeping some people out, while in neighborhoods where the original families have moved out, leaving their homes to deteriorate or be remodeled, newly arrived affluent gentrifying homeowners use preservation to reform and convert their neighbors and justify or legitimize their own presence. Both groups may feel a sense of exclusivity and entitlement in protecting their private residential landscapes, but long-term elites do not have the same burden as gentrifiers, the newly arrived preservationist homeowners, in constructing a conceptual historic community. Elite family members perform their social identities within the context of homes their families have owned for years, they recount narratives of community and shared memories of collective celebrations and invented traditions in the neighborhood place. They transform neighborhood materialities into cultural capital that communicates a social identity and position that is implicitly exclusionary. In these cases, exclusion may be more important than any aesthetic values associated with the historic preservation cosmology, except to the extent that legislation tends to keep things the same—as they have always been. "Preserving the 'look of the landscape' is the primary intention" (Duncan and Duncan 2001: 390). Elites focus on keeping the appearances of houses and neighborhood the same, and seem to be threatened by encroaching activities at the edges of their neighborhoods more so than by the potential destruction of an historic monument or neighborhood at some distance from themselves.

By contrast, the gentrifiers are engaged in a "reterritorialization" of a neighborhood and seek to uncover and restore the original architectural features of houses. They are like the new middle classes Neil Smith describes moving into the "frontier" of deteriorated urban centers, whose identities are constructed by trendy forms of consumption to differentiate themselves from the old middle classes (1996: 114). In contrast to the relatively long-settled elites, these new middle-class homeowners acquire properties in neighborhoods among ethnically and class-diverse residents where they initially constitute a minority professing historic preservation cosmology as an aesthetic norm. As a result, they are much more conscious of the differences between themselves and their neighbors, they are more vocal and they are more active in promoting their values. Seeking and finding like-minded, preservation-oriented homeowners for mutual support, and for sharing knowledge and experiences are critical to their survival in the new neighborhood. Collectively, these homeowners are challenged with constructing an historic imaginary for not only their houses and also for their neighborhood and beyond. Constructing the imaginary links together a community of residents who share the same aesthetic and historic values and vision for the future, and who share in the desire to realize and activate the same moral project. Surrounded by homeowners whose aesthetic preferences so obviously contrast, they seek out fellow believers that can collaborate in producing the historical landscape ideal.

Fear of other aesthetic values, then, motivates preservationist homeowners to prioritize the promotion of education in their own neighborhoods and the city at large. Preservation organizations, established in one form or another in all the cities reviewed here, typically run home tours, but also provide workshops, lectures, printed material and organize other local events to raise awareness of historic preservation values and practices. Supporting their membership with resources and knowledge about restoration practices also raises member awareness. Additional services sometimes include researching individual house histories, offering awards programs, and distributing monetary grants to help defray costs of remodeling. Advocacy in the neighborhood includes "policing," reporting suspected violations of preservation regulations to the city, as well as working with elected officials and city building and planning departments to secure better oversight. The most visible form of advocacy includes showing up at city hall to make voices of support for, or in opposition to, particular projects heard, and it may also include members serving on design review boards or historic preservation commissions.

Preservation organizations aim their educational or advocacy efforts to include local residents of different populations whose aesthetic preferences may pose a threat to the historic ideal. Many preservation homeowners complain about inappropriate remodels such as stuccoing over shingles and wood siding or replacing wood windows with aluminum or vinyl replacements. Homeowners who engage in these types of remodels and repairs are often long-term residents or more recent arrivals who simply want to "modernize" and update their old houses. Preservation activists assume that many of these homeowners simply lack the knowledge about restoration practices that would preserve what they perceive as the highly valuable original features of the house. While some homeowners appreciate the information, many more seem uninterested, or are defiant about modernizing their homes, claiming they have the right as property owners to do as they please. Landlords, also, are not much interested in historic preservation as they try to maintain properties in the least expensive manner possible. Recent immigrants or members of ethnic groups who express different cultural preferences and tastes may employ remodeling strategies that seek to transform a small bungalow into a midcentury tract house or a mansion. Another category of homeowners includes those who buy run down houses to quickly remodel and resell. Sometimes these owners are interested in superficial cosmetic repairs that appear historically sympathetic, but too often serious problems with house's structure, and electrical, water, sewage, and heating systems are overlooked in order to speed up the resale. Some are not homeowners at all, but are renters who have found affordable housing in an aging neighborhood and whose daily lifestyles contribute to further deterioration. Lastly, the most reviled of all are the developers whose main interest is to demolish old houses and build brand new single- or multiple-family houses in their place.

The descriptions of individual cities that follow divide the discussion into two related chapters. The first focuses on two cities, Riverside and Ontario, where elites have been able to make historic preservation legislation work for them, while the second examines gentrification in the two cities of Monrovia and Pasadena. In the former two cities, with stable elite communities whose existence preceded the establishment of the historic districts where they live, preservation laws and policies are appropriated and embraced to reinforce the exclusionary boundaries of the district community. Exclusion sentiments may be focused within the neighborhood towards newcomers, or at protecting its edges. District residents, however, only weakly adhere to or promote preservation cosmology, although there are many passionate preservationists scattered throughout each city and its neighborhoods. While both cities have citywide preservation organizations, the cosmology aimed at retaining and restoring architecture to its original conditions seems to operate inconsistently on the individual homeowner level. Rather, elites embrace their heritage as a means to keep everything the same—materially, but also socially—to preserve the notion of an enduring social order.

The analysis examines how each locality has succeeded, or failed, at unifying members of its organization or association, or the neighborhood into some type of preservation community, a social-symbolic community produced and reproduced with varying degrees of preservation concerns. Of particular consequence are the programs and membership interests of local preservation and neighborhood organizations in localities where this study has focused. An exploration of some depth reveals the ways preservation and neighborhood organizations interact with city government, and how the issues each focuses on are at least in part conditioned by how each city's preservation legislation has been historically structured, as described in the preceding chapter. Of special note are the relationships particular neighborhoods or citywide organizations maintain with the city, especially with city council, the planning departments and historic preservation offices, and historic preservation commissions or committees that review construction proposals from citizens.

City of Riverside

Riverside and Ontario are two localities marked by fairly weak expressions of preservation cosmology. This is not to say that passionate historic preservationists and purist restorers cannot be found among homeowners or consultants and contractors. Rather, one is likely to find the strongest feelings among members concentrated in the citywide historic preservation organizations even though their members are drawn from districts and areas throughout the city and not from any specific neighborhood. Moreover, while these preservation organizations may have some influence with city hall when contentious issues are debated, other organizations and individuals may have even more influence on particular historic or cultural issues. The lack of any singular community power to advocate on preservation issues may be due to the fact that both cities have fairly well developed and professional municipal historic preservation programs that handle most applications and complaints. As noted in the previous chapter, the city of Riverside led the region in establishing landmarks legislation in 1969, while the city of Ontario attempted legislation in the early 1990s. But while Riverside's governmental approach seems to have fostered the growth of two active citywide preservation organizations and a substantial professional infrastructure of expert consultants and contractors to obtain approvals for applications and proposals, Ontario's governmental approach has seemingly co-opted its sole preservation organization and discouraged its attempts to salvage a disappearing urban core or to generate much local enthusiasm.

The different historical dynamics in Riverside that successfully focused preservation concerns within the city's museum and museum staff, and produced the city's Cultural Heritage Committee to review historic preservation projects, also helped foster the creation of a citywide preservation organization. The Old Riverside Foundation (ORF) met for several years informally before it was incorporated as a non-profit in 1979, and a secondary group, the Riverside Renovators, split off several years later after disagreements with ORF's leadership, but never formalized its organization by incorporating. Ontario's sole preservation organization, Ontario Heritage, was originally established in 1976 under the name of the Ontario Historic Landmarks Society, which was changed in 1992 to Ontario Historical Society, and then in 1998 to its current moniker.

The Old Riverside Foundation is a widely recognized institution in southern California having pioneered many educational and social programs promoting preservation over three decades. Much of ORF's energy in the early days was focused on the intense redevelopment activities centered in Riverside's downtown area, which continues. Developers' proposals for modernizing the civic center required demolishing many older commercial and residential buildings. Postwar development concentrated on urban revitalization efforts, such as replacing the old library with a modern one, or building new housing tracts in former citrus groves. One of the most contentious issues involved the threatened demolition of the Mission Inn (1876), considered by many as the heart and soul of historic Riverside. The broad scale of proposed construction efforts dismayed some of Riverside's elites who feared the destruction of the city's history and architectural heritage. These residents, which included UCR professors, rallied to advocate for protecting old buildings from demolition. While the group was ultimately successful in saving the Mission Inn, they also lost a series of battles with the city council that favored the developers' interests for downtown development.

The Old Riverside Foundation was particularly concerned about the fate of large Victorian homes and civic buildings in the central city, and also "grove" houses built on dispersed agricultural sites within the city's borders. According to a former ORF president, in the 1960s and 1970s developers pressured many older homeowners to sell their Victorian houses and surrounding orchards to make way for the construction of new housing tracts. Although ORF lost many early battles to save these buildings, when demolitions occurred the organization was able to salvage building materials, doors, windows and fixtures that could be sold to residents engaged in their own restoration projects. ORF still warehouses the materials in its headquarters in the Weber House. Another significant program developed in the early years, which continues into the present, is the relocation of threatened Victorian homes, especially old grove houses. The city "sells" the houses for $1 contingent on the buyer acquiring a lot on which to relocate and restore the house, and ORF promotes these sales on its website. Many houses have been moved to empty lots in existing historic districts, although this practice has been criticized because it places too much emphasis on saving the "object" rather than retaining the physical context of the house.

As ORF became more active in the 1980s, some of its most active advocates became members of the city's Cultural Heritage Board; in fact some residents say that for about ten years the membership of the board was virtually identical to ORF's leadership. Neither board members nor museum staff were necessarily trained in architecture history or historic preservation, but learned through their advocacy and hands-on experiences. ORF favored saving Victorian buildings in the 1970s, perhaps because the fifty-year minimum for an historic landmark designation focused most of their attention on buildings dating from before 1920. In Riverside, Victorians were also seen as having intense symbolic meaning and broad civic value because they served as significant and tangible markers of the city's early agricultural prominence and its role in the origins of southern California's citrus industry. Many of the earliest landowners and citrus growers built these ornate houses as their primary residences. One of the biases resulting from this focus on preserving the very elegant and ornately decorated Victorians was a devaluing of other vernacular buildings and architectural styles that was not acknowledged until later as the city's historic preservation program evolved.

As the city's preservation program matured and became more professionalized, and the Cultural Heritage Board moved into the municipal planning department in the 1990s, ORF's own organization began to change as they intensified their educational efforts. One former member asserted that the reason why the city moved preservation activity to city hall was because the Board had become too powerful under ORF's influence, and complained that now the city council ignores them. On the other hand, current ORF leadership includes several planners and government employees as well as one realtor, and their advocacy tactics now include occasionally filing lawsuits to compel compliance to city preservation laws. On the cultural side, from their earliest days ORF sponsored a 12th Night Celebration, a progressive dinner, on January 6 each year. Recalling Victorian England of the 1870s, participants dress in period costumes and some hire horse-drawn carriages to take them from house to house to celebrate. ORF has also run annual home tours recruiting a variety of houses from different historic districts throughout the city. One of the significant characteristics of the ORF is the extent it promotes appreciation for the actual history of the city and region. Special events, guided walking tours and visits to local landmark sites, led by historians and other professionals, emphasize historical and cultural details as part of the educational mission. ORF also holds an annual awards program to acknowledge exceptional examples of restoration and maintenance of historic architecture.

While ORF operates as a 501 (c)(3), has an elected board of nine directors and a wide range of established programs, the Riverside Renovators operates as an informal social group that gathers each month at a different site. Usually a member of the the group offers their house, typically under renovation, as a meeting site. At these monthly events, participants are provided with a guided tour and explanation of some of the owners' restoration plans. Alternatively, an historic house that is on the market might also be used as a meeting place where the realtor can explain its historic physical and social attributes. While the Renovators are interested in historic preservation, they appear to include many more contractors, realtors, collectors, and those who like to tackle renovation projects, while ORF members seem to include those more interested in history and cultural heritage. The Renovator group seems particularly interested in ridding houses of modernizations and salvaging original fixtures and materials from older houses to achieve an authentic restoration. The purpose of their gatherings, however, seems to be primarily social, centered on potluck dinners and plenty of refreshments, while also fostering connections between homeowners and restoration professionals.

At the local neighborhood level of Oak Streets, residents are quite familiar with ORF and their activities, and some residents have agreed to have their homes put on the home tour. Most residents, however, are not active members in the organization, although they do know about the salvage and other programs. With the exception of one resident, a designer who had an elaborated concept of her house "embracing her," most local residents have not declared a passionate interest in or an awareness of a strict interpretation of restoring their homes to their original conditions. Even so, contractors employed by these homeowners to do renovations are fully aware of and engaged in restoration ideals and practices. Interviews with some of the residents revealed that while they seem genuinely interested in "restoring" their houses, having modern conveniences and remodeling to expand or reconfigure spaces for a more comfortable contemporary lifestyle takes precedence over purist notions of retaining the original materials and built features. Some residents also acknowledge knowing about the Riverside Renovators, what one resident called "purists," but no one interviewed in this study was an active participant.

There is, however, a neighborhood association operating in Oak Streets as there are in other historic districts. Established in 1997, as an official city Neighborhood Association, the Historic Oak Streets Association formed in response to a problem with the students of nearby Riverside Community College (RCC) parking on residential streets during the school day where parking was limited to one hour. The issue was exacerbated by the planned construction of a new digital library at the college, which eliminated much of the student parking. One of the Oak Streets' residents called the police and the college to see what could be done about the issue. Realizing that the neighbors had not been notified that RCC was planning to build the library, and that the loss of parking would have significant impacts on the neighborhood, he and others organized a neighborhood meeting. At the meeting one of the attendees suggested that the group form a Neighborhood Association in conjunction with the city's Office of Neighborhoods.

The resulting Historic Oak Streets Association today represents about 300 households in the historic district plus two additional streets at its southern border, added in 2004. It is important to note that the association inverts the word "historic" placing it at the beginning of the name rather than adapting its official district designation, Oak Streets Historic District. As two board members explained, the neighborhood association is affiliated with the Riverside Neighborhood Partnership, which evolved out of earlier Neighborhood Watch programs. The city's goal is to link every neighborhood to the city to coordinate the flow of information and encourage disaster preparedness, crime fighting, beautification, and social outreach. Since 2003, the office has been conducting leadership training to those who want it and provides opportunities to network with other neighborhood representatives, city officials, and business people in the Neighborhood Council. As noted in the previous chapter, the city would like the neighborhood association to represent all of the Oak Streets Historic District and NCA, but the association's current leadership has so far resisted. They have said, among other things, that the size of the NCA is too large to maintain the personal relationships necessary for the hand-delivered newsletters to every household. They prefer maintaining the smaller historic district, and so they think it unfeasible to expand the membership.

The primary purpose of the association is not historic preservation education or advocacy at the neighborhood or city level. Realistically, the city already has two preservation organizations focused on that task. Among the historic districts, only Mount Rubidoux has adopted residential design guidelines for its historic houses, written largely under the guidance of one of its own residents, the former state historic preservation officer. Mt. Rubidoux's neighbor, Colony Heights, is also reported to be quite active in promoting preservation cosmology in the city. Both of these districts are geographically proximate to the downtown area. Rather, the mission of the HOSA vaguely promotes disaster preparedness, crime prevention, and general social welfare, reflecting the city's priorities:

> to speak in a unified voice in addressing concerns and maintaining a high quality of life in this neighborhood of unique and historic homes. Collectively, we will strive to maintain a communication network to ensure a safe and secure neighborhood to care for each other and our homes. (*Oak Streets Quarterly* 9 (2) (2010): 2)

According to some newer Oak Streets residents, association members seem overly preoccupied with crime, or potentially criminal behavior; the leadership coordinates with the police department and sends out e-mails whenever suspicious activity such as loitering, theft or vandalism is spotted. One leader said she considers it her responsibility to "know who belongs" and to notice when someone or something is out of place. It is important to note that the two primary leaders had recently moved into the neighborhood from outside the city of Riverside.

A recently arrived homeowner, however, commented that some long-term residents and leadership seemed unusually concerned about low-income minority residents who live in rental housing along the eastern edge of the district, and who move through the neighborhood to access transportation and shopping. These renters used to visit a small "mom-and-pop" convenience store, the Kawa Market, along one of the neighborhood's main streets at the western end. The owners of the store were immigrants from China who had bought the property in 1991, and continued to operate the same type of business as the previous owner. As early as 1995, Oak Streets neighbors began complaining to city officials about the market attracting the "wrong" element such as Blacks, Latinos, and Asians. Although the original market predated city ordinances dealing with "problem properties," the city eventually found a solution. In 2007, in its capacity as the Redevelopment Agency Board, the City Council voted to eliminate the market by offering to purchase the property from the couple for a set price based on their own appraisal. The city made it clear to the owners that if they did not accept the offer, they would be subject to eminent domain proceedings. Supporters of the market reacted to the city's maneuvering as unscrupulous, which created a very public backlash and contributed to the election of a new city councilman shortly after. Memories of the Kawa Market incident continue to provide a touchstone for discussions about ethnic and class issues in the neighborhood.

Although the Historic Oak Streets Association is not necessarily concerned with promoting historic preservation, residents are very aware that they live in a designated historic district and residents of the NCA do not. Some residents mention that they have worked with the city's Historic Preservation Officer to obtain landmark or structure of merit designations for their homes, or have consulted with her or very experienced contractors about design issues in "properly" restoring their houses. An association leader said that she checks with the Historic Preservation Officer whenever she sees construction in the neighborhood to make sure that homeowners have permits for renovations. She noted the ongoing remodeling of a nearby house, complaining that perhaps the vinyl windows were not exactly appropriate, and suggesting that vigilance was required because there are "loopholes" in approvals issued between the Building Department and Cultural Heritage Board. But beyond making sure projects conform to zoning requirements such as fence heights and setbacks, she expressed little knowledge about or appreciation for stylistic integrity, nor much interest in promoting the original architectural character.

The resistance that some residents, especially long-term residents, express about making the district more inclusive may stem from the fact that the six and one-half blocks that constitute the Oak Streets Historic District was a strongly integrated neighborhood before the district was officially formed in 1988. Several residents described one of the streets as central to neighborhood Christmas celebrations that also included a progressive dinner. Another noted that neighbors collaborated in writing down oral histories of older residents. Older residents especially recall block parties, BBQs, 4th of July celebrations and Halloween festivities that brought families together, often through their children, for intense socializing. Many of these social attributes, rather than architectural features, give the neighborhood its special "community" quality, which some newly arrived residents say especially attracted them. One couple moved from Orange County to retire in the Oak Streets because they had always wanted to live in a "real community" while another mentioned that they had moved in from an automobile dominated upwardly mobile suburban neighborhood so they could "walk on sidewalks."

Although many residents of the historic district were clearly attracted to the architectural consistency of its houses and a sense of community, a former city planner described a "pride of heritage" was the more dominant sentiment in the city than a commitment to preservation. To several residents "historic" meant that people recognize the "historic values" of their houses and find them worth preserving for their own sake, not necessarily remodeling them to make a status statement. Another said the neighborhood reminded her of a small 1950s' community, suggesting nostalgia, that made the houses worth preserving, while yet another resident mentioned that the neighborhood was historic because people are able to live in houses associated with the official history of Riverside.

Still, exclusionary sentiments dominate decisions about what streets, houses or people to include in the "community" of Oak Streets Historic District, which tends to focus on boundary maintenance. Longer-term older residents argue that houses in the NCA are of lesser quality than those of the historic district and have been remodeled in inappropriate ways that do not merit recognition as historic resources. Or, they offer, the NCA is a neighborhood where older houses are interspersed with more modern "ranch" style houses. These residents point out the people who live in the NCA may also include "the wrong type" of people who are working class, or lack the culture and taste commensurate with that of historic district residents. Their view of the 2,700 houses "on the other side" tends to be one of suspicion and fear in contrast to the intensely social neighborhood with which they are most familiar. Newly arrived residents in the historic district have suggested that older residents' sentiments suggested a racial and class bias and that preservation desires, to the extent they are present, were focused more on maintaining the historic socioeconomic order of the neighborhood. These sentiments can be seen in the recent conflict over the Kawa Market. Of course, residents who live in the NCA do not necessarily understand what historic district residents hold against them, even when their friends who live there try to explain.

Several residents mentioned that "some people have too much power" in the Oak Streets Historic District. On a few occasions, current and past association leaders professed that their group "is not political," by which they meant that the association does not advocate at city hall on matters concerning historic preservation or other issues. One long-term resident, however, suggested that the neighbors are "very political" and mentioned residents' ties with local officials and other power brokers. Residents do make their feelings known to their city councilman with whom they maintain regular contact, and the police department and Neighborhood Council, but this is not what residents complain about. What they mean is that some old-time residents seem to have more influence at city hall than newer residents, and in ways that sometimes seem unfair, and that exclusionary tactics are applied to neighbors within the historic district itself.

One such example is found in a remodeling conflict involving a couple who in 1991 had bought a one-story 1929 Mediterranean Revival house, which had been designated a structure of merit. Soon after the homeowners decided that their family of four needed more room. The couple hired an architect to design a second-floor addition after determining that the back yard did not provide enough space for an extension. After more than five design iterations, and with particular sensitivity to minimizing the appearance of the second floor addition, the architect presented the proposal to the Cultural Heritage Board and won project approval in a close 4–3 vote. Within two weeks, three of the couple's neighbors—one immediate neighbor, one who lived across the street, and another living down the block—filed an appeal with the mayor and city council. The appeal argued that the design was inconsistent with the original historic character of the house and noted that all two-story houses in the neighborhood were original. Moreover, it suggested that permitting the second story, which they saw as clearly not a "minor alteration," would set a precedent and threaten the "integrity" of the historic neighborhood.

Appeals of Cultural Heritage Board decisions are turned over to the recently formed "Land Use Committee," whose members are drawn from among members of the city council. Although the planning department wrote a letter of support for the original Board decision, the Land Use Committee approved the appeal. The homeowners had prepared for the appeal hearing by researching all the second story houses in the neighborhood and securing the signatures of all the neighbors (a considerable number), except those opposed, on a petition supporting the project, which they submitted before the meeting. According to the couple, the Committee suggested that they "should have asked all the neighbors first before proceeding with the project." Moreover, the Committee seemed to ignore the petition with signatures of a majority of the neighbors. Although the couple presented evidence of neighborhood support and second story precedents, none of the evidence appeared in the city's documentation on the case. After they were turned down, they decided to abandon the project. The couple speculates that one of the opposing neighbors was well connected with council members having worked in local government and volunteered on city commissions for years. They believe that these connections gave their opponents an unfair advantage in the decision. After all, they said, it was impossible to find a written requirement that the "homeowner seek approval from the neighbors" in advance as a condition for approving any construction in an historic district.

Residents of the Oak Streets Historic District and NCA are unlikely to become more proactive in advocating and educating people about historic preservation. One reason is that most of the building stock in both areas is newer and has not deteriorated over the years, thus requiring less dramatic interventions in order to restore the original conditions. Also, residents generally seem to have a much less purist approach or are more flexible in their remodeling practices. Despite what historic district residents say, the NCA has been home to far fewer lower income residents and renters than they realize and has been attracting younger families who appreciate the historic charm of the neighborhood regardless of the district designation.

The most important question about unity in the historic district revolves around the social divisions between the two geographic areas based on perceptions of superiority by one rather exclusive historic section in contrast to the rest of the NCA. While the exclusionary motives of historic district residents may be focused on elitist factors such as higher incomes and social standing, at least some of the principles they use to justify their claim of entitlement to a separate status comes from the city's own historic preservation laws and policies. Specifically, the distinction described in Title 20 between the NCA and Historic District involves the ostensible "lesser" significance or concentration of such housing in the former area compared to the latter. And, although the city has long since abandoned this distinction as it pertains to material qualities, it has been popularly interpreted to have a social meaning. Even the initial designation of the historic district did not follow the codified description, but used the willingness and desire of a socially exclusive group to present themselves as the sole historic district and thereby acquire increased status, recognition and social legitimacy that has endured since they first asserted their claim of distinction.

In recent years, as many younger families have bought houses in the historic district and NCA, sentiments about inclusion have begun to take hold. In 2009, a group of residents organized a new "green" organization to promote sustainable living and lifestyles in the neighborhood and the city. And, in the summer of 2010, another group of residents began meeting with the city councilman about issues of neighborhood safety and formed another neighborhood association called Neighbors of Oak Streets to involve and represent residents of the whole NCA. Leaders from both of these organizations signed up for the city's leadership-training program, and participate in the city's neighborhood council program. These new groups welcome residents from the entire neighborhood, including the historic district, and all three groups, including the Historic Oak Streets Association, all cross-list their programs and activities.

Ontario's diffuse preservation efforts

As one of two cities with diffusely focused historic preservation sentiments, the city of Ontario has only one citywide education and advocacy organization, Ontario Heritage, which took its current name in 1998. As noted earlier, Ontario's efforts at establishing historic preservation legislation began in the 1990s with a citizen's commission that was soon disbanded after proposing a legislative program the city considered too restrictive on developers. In 1993, the city adopted much watered-down legislation and, in 2003, revised its laws. The role of Ontario Heritage, which only came into being after the city had adopted the weaker legislation, has itself been inconsistent and weak, but there are no other organizations representing individual preservationists or the six historic districts in the city. Ontario Heritage has hosted occasional home and cemetery tours, and benefit dinners, but it has also had persistent problems in attracting members and leaders who can commit to running the organization. In fact, much of the material and logistical support for the organization has been provided by the city's planning department, which prints flyers and aids in coordinating events for the organization. As one planner referring to educational outreach and fundraising lamented, "If we don't do it, they don't do it."

Many of the homeowners in Ontario interviewed for this study made very positive comments about the historic preservation programs managed by the city's planning department. In addition, Ontario's public library houses a large collection of photographs, newspaper articles, and historical publications documenting the city's history and its architectural heritage. Homeowners of historic houses often consult with the archives in the library's Model Colony Room to learn about their houses and prepare proposals for landmark designations. Homeowners who have undertaken renovations say they have received planning department support in obtaining approvals and consider the city's programs "extremely good" and pro-preservation. Project reviews by the Historic Preservation Commission have been favorable and several homeowners commented that the permitting process is fairly lenient and allows wide interpretation of the design guidelines. A few homeowners, however, also say that code enforcement is quite strict, and a number complain about the downtown development plans that include the continuing demolition of historic buildings. They have singled out the Housing Department that buys up individual properties to sell to developers for new construction.

The designation of historic districts has followed along the same lines as Riverside. Using prior historic resource survey data, the city identifies areas with consistent residential architecture for possible district designation. The city holds meetings with residents of a proposed district, and, once agreement of a majority of residents is secured, prepares the designation proposal for submission to the Historic Preservation Commission and city council. In one recent case, a group of residents from one neighborhood came to the city council to ask if they could become a district, but most districts have been formed because of the city's initiative. City planners have put forth great effort to develop the historic preservation program that is noted for its professional accomplishments and stays current with goals and objectives of the state office of historic preservation.

The initial activities of Ontario Heritage during the late 1990s and early 2000s seemed to promise that Ontario would have an active community organization to promote and advocate for historic preservation. In 2003, the owners of a 1909 two-story Craftsman house near the civic center decided to upgrade the exterior and cover the wood siding with stucco. The couple bought materials and had begun attaching the lath to the sides of the house when a neighbor asked the wife if they had obtained a permit from the city. The next day the wife went to the building department to get the permit, but discovered that her house was considered an "historic" property. The city had recently implemented a new computerized database based on the original 1983 historic resource survey. The city refused to issue a permit and code enforcement put a "stop work" order into effect. The couple sought relief by appealing to the Historic Preservation Commission, which also turned down their request to stucco their house, or to de-list it from the database of historic properties. The couple appealed the decision to the entire Planning Commission and was represented by an attorney who complained that the homeowner did not know the property was on a list when it was purchased. In asking for the house to be de-listed, the attorney referenced other homeowners who had been successful in requesting the same action.

Two members of Ontario Heritage appeared at the appeal hearing to express their views. One member suggested that the nearly original condition of any historic house has great value and homeowners should work to maintain the "historic integrity of the house and neighborhood." Another member argued that stucco applied to wood siding could cause "wood damage, rot and mold" and would incur "structural harm" to the property. Two of the couple's neighbors, however, said they would like to see the neighborhood cleaned up and wanted to put stucco on their houses as well. In the discussion among commissioners and the planning director that followed, the conversation focused on the survey's "list" of historic resources, which for all intents and purposes remains a virtual secret until a homeowner seeks a permit. Some commissioners wondered whose responsibility it was to notify homeowners they were on the list, and one worried that the list "stigmatizes" homeowners because they are unaware. The planning director thought it was a responsibility of a real estate agent to advise the prospective homeowner, and that the city's historic preservation staff was too underfunded to enable them to accomplish this task. Finally, the conversation turned to the economic value of the historic house that would be lost if they permitted the stucco project, and could even result in "blight." The commission decided against the appeal and ordered the couple to remove the nails and lath, and restore and paint the original wood siding. As a consolation prize, the Planning Department awarded the couple a certificate of appreciation for their efforts to comply with historic preservation requirements.

While advocacy for residential projects has been one of the main accomplishments of Ontario Heritage, advocacy for civic buildings has been less visible. Several homeowners interviewed in this study said they had joined Ontario Heritage with expectations that the organization would be a strong advocate for preservation issues at city hall, especially where downtown development was concerned. Within a short period of time, however, they dropped out of the organization because, they said, they were disappointed with the organization's inaction. Some suggested that there were too many conflicts among some members of the board, while others noted the organization seemed misguided about preservation. As a way to underscore the last point, members pointed to the house owned by Ontario Heritage's former president and mayor, which he built new as an "authentic reproduction" of a Victorian house. To understand why there was a mismatch in goals and vision between Ontario Heritage's leadership and some of its members, an analysis of the board leadership shows how the overlap with city commissioners and elected politicians may have had some influence. Planning and Historic Preservation Commissioners have been leaders on the Ontario Heritage board, and one former board member is currently on the city council. This overlap in membership tends to put brakes on any discussion of preservation projects that would threaten developer interests, including the preservation and restoration of unreinforced brick buildings, which had dominated the downtown.

Some former members of Ontario Heritage call the organization just flat out dysfunctional with shifting leadership that has had trouble setting goals and developing a comprehensive strategy, not to mention recruiting a viable, independent board. One former member who also served on the first historic preservation commission claims that the problem with Ontario Heritage is really a problem with the city council, which has a long history of involvement with all kinds of development issues in the city. She suggested that one twenty-year old problem was that all the council members came from northwest Ontario, from the historic district called Armsley Square. As property owning elites, their primary interests were in the eastside development of the airport, shopping malls and residences on land some of them owned. The council was interested in making money and was very pro-business, while ignoring quality of life issues throughout the rest of the city. The neglect of the area south of the railroad tracks, including the lack of oversight of the 1,200 previously "historic" houses that had to be de-listed from the 1983 survey, was partly a function of diverted attentions at city hall.

In fact, this preservationist claimed that the city council did not want to do preservation because, as proponents of property rights, they did not want "to be telling people what to do." She had seen this sentiment before. In the early 1990s, she had been approached by a number of Ontario businessmen who, with the support of the local Chamber of Commerce, asked her to help them apply for a grant for a pilot Main Street Program for mid-sized cities under the auspices of the National Trust. She had previously created a similar program in South Carolina, where her husband had been stationed in the military. After preparing the application for the National Trust grant, the proposal went to the city council for endorsement, but she said the council refused to approve it.

Like other cities with historic preservation programs, the residents of Ontario expect the city to monitor or police the districts and prevent inappropriate remodeling. This is one of the functions of the much disliked code enforcement division. By far the most frequent complaints come from the residents of Armsley Square Historic District, about 3–5 per year, but no one could have anticipated that in 2008, one of the neighbors would strip fixtures from her tony historic home and put them up for sale on eBay. Even more surprising was the fact that she had been an active member of Ontario Heritage, and had received a Mills Act contract reducing her property taxes in exchange for investments made in restoring the historic home. Apparently, her house had fallen into foreclosure and, failing to get a fixed-rate loan, the homeowner wanted to "cut her losses," and so she began stripping fixtures. As in many historic neighborhoods, however, it is difficult to escape the watchful eyes of neighbors. Code enforcement responded by changing the locks to prevent the owner from removing and selling more items. Eventually, the sales of the fixtures on eBay appeared in the local newspaper and became the scandal talk of the town.

Conclusion

Preservation experiences in Riverside and Ontario suggest that elite homeowners have engaged their local municipal agencies in historic preservation protections, not so much out of a passion for preserving material history but as an exclusionary technique. Although sections of both cities have experienced deterioration in particular neighborhoods, especially in lower income areas, elite residents have embraced historic preservation primarily to protect their own neighborhoods from perceived threats from developers or newcomers whose housing values vary with their own. When a newcomer proposes to build a second story, even if it is sensitively designed, or a homeowner strips away the original fixtures in the house, elite sensibilities are offended. These cases reveal that claims about threats to the "original" character express a "pride of heritage" aimed more at preserving the social and economic order rather than the neighborhood's material conditions. Elite neighborhoods in both cities are filled with houses, which, for the most part, have been well maintained because their owners had the resources to keep up the properties. Although some houses may have suffered from "modernizations," new owners are able to restore and enhance the original qualities through much appreciated remodeling. Homeowner preservation activities are constrained and enabled by neighborhood norms regarding the aesthetics and materiality of housing.

The Gentry Move In: Education, Reform, and Advocacy

Unlike neighborhoods in Riverside and Ontario where elites have maintained a constant presence and effective control over preservation efforts in residential areas, the cities of Monrovia and Pasadena have, since the 1980s, initiated preservation and experienced gentrification in what were some of their most deteriorated historic neighborhoods. These are both cities where advocacy has become quite strong, but also where a social movement aimed at re-territorializing and transforming the material character of the city's residential neighborhoods continues to take place. Behind these efforts are some professionals, but it is mainly residents who have acquired on their own, and collectively, a sufficient level of sophistication and passion to promote, even proselytize, the preservation cause. Monrovia and Pasadena residents distinguish themselves in the construction of a local symbolic community committed to historic preservation advocacy, even if the unintended consequences are exclusionary. Homeowners in both cities have restored dilapidated houses in neighborhoods filled with residents who have different aesthetic ideas and have learned that persuading the neighbors and maintaining constant vigilance are both necessary. Their community is constructed around the idea of history as a place-based social imaginary, which acts to identify its individual members and define their commitment to a moral project of restoring architecture and neighborhoods to their original conditions.

Constructing the preservation community requires the imagination of an idealized historic landscape full of individually restored houses inhabited by people who enjoy turning back the clock, if not to relive the past, at least to showcase it. Implicit in this construct is that the houses, and the neighborhood as a whole, instantiate a scenario of prewar middle-class family life embodied in perfectly restored material conditions. It may be a museum, but it is a living one. Constructing the community depends on creating a narrative of place that relies on historical facts and aesthetic imaginings. Community is experienced as a "structure of feeling" in which period architecture provides the symbolic retreat to the past (Tilley 2006: 14). Promoting the establishment of historic preservation regulations, or the creation of an historic district, requires advocates to assertively carve out the mental or cognitive space for an historic vision, one that is pragmatic enough but also suggestive about resolving some set of socioeconomic issues (Wilson 2004: 44). Residential preservation arguments often promise more than just the material improvement of older rundown neighborhoods to eliminate. They also suggest social improvements in the quality of life, street safety, lower crime rates, and stabilization of property values; all conditions stereotypically associated with lower income families. In essence, the preservation narrative implies that a place needs cleansing, the repair of old houses and removal or displacement of populations who live in them (Herzfeld 2006: 136).

The core idea in constructing the historic preservation narrative rests with the intention of returning houses and neighborhoods to their original material conditions, suggesting a more wholesome, idyllic community undisturbed by the problems of contemporary society. The politics of preservation, then, is largely a grassroots campaign that depends on a substantial number of local supporters and converts. Advocates must win over local politicians and city hall, but they must also convert their neighbors and engage them in the historic imaginary. They emphasize the "look of the place" in their meetings and negotiations with public officials to accentuate the centrality of aesthetic properties to the proper representation of local history (Duncan and Duncan 2004). Their emphasis promotes the nobility of aesthetic regulations in conceptualizing and possibly marketing the city's identity, even as the regulations act to exclude houses and their occupants that do not conform.

The space created by the preservation narrative constructs house rehabilitation strategies as the entry point into membership in the community, which preservationists interpret to entail a series of moral obligations shaped by an aestheticized history. The restoration project argues that when one buys an historic house or a house in an historic district, one is buying a piece of history and with it come duties and responsibilities for its care. At minimum, these responsibilities are spelled out in the city's historic preservation requirements, but there is a larger moral sphere in the neighborhood that envelops and constrains homeowners. The preservationist notion of responsibility runs counter to a private property rights argument some homeowners assert to justify doing to their houses whatever they want. Rather, the preservation project requires a different set of attitudes and behaviors oriented at respecting, even submitting to, the idea of valuing the original material conditions and rehabilitating them over the long term.

Restoration beliefs and practices construct the preservationists' moral community. Whether a house is part of a historic neighborhood, or stands on its own, the homeowner's duty is to restore and protect the original features of the property. This vision of an idealized historic landscape also requires adherents of the cosmology to educate neighbors and newcomers to the neighborhood about the history of the place, and to encourage all homeowners in the historic community to realize their responsibilities and commitment to its restoration, and possibly recruit them to advocacy. Unlike Riverside and Ontario, neighbors who have strong expectations about advancing what they see as a morally compelling cause seek to persuade new homeowners. Here, education is not just about passing on information, but reforming homeowners' sense of themselves, their relations to their homes, and what they owe to the community.

City of Monrovia

Although the city of Monrovia only established its historic preservation laws in 1995, preservation-oriented homeowners have had a strong presence in the small city of 39,000 since the 1980s. In 1980, the Monrovia Old House Preservation Group (MOHPG), the city's sole preservation organization, was founded by two married couples who had recently moved to Monrovia, and were rehabilitating their early twentieth-century homes. As one of the founders, a college professor, tells it, he and his wife were taking a walk one evening, when they happened upon another homeowner who was doing work on the exterior of his house. They struck up a conversation, which led them to form a friendship based on shared interests in restoring old houses. Within a short time they decided to look for other residents who had also recently bought old houses in Monrovia and were seeking social support for and practical knowledge they could apply to their restoration activities. Soon, they began meeting in one another's houses, much like the Riverside Renovators still do, so they could show off their handiwork and engage others in discussions about it. Initially, MOHPG met under the auspices of the Monrovia Historical Society, which had formed the previous year to assume ownership of a locally significant Victorian house and turn it into a museum.

Early members of MOHPG promoted historic preservation of Monrovia's prewar houses, especially Victorian and Craftsman houses, by delivering flyers door-to-door and making presentations at local clubs and organizations. The participation of new Monrovia residents grew, but there was still resistance to preservation concepts among long-term residents and business interests, including several real estate developers, who desired new construction to revitalize the city's core. In a small town, these sentiments were clearly felt and understood, which is why the early preservationists focused exclusively on restoring old houses and named their organization accordingly. The founder of the group explained that "MOHPG" was an ambiguous acronym because it could also be read as MOnrovia Historic Preservation Group—which is now its actual name.

According to early MOHPG participants, the city of Monrovia had a very "bad" reputation in the 1970s; it was a dangerous place. It was known for gangs, drugs, and even "race riots" at Monrovia High School. Many of its old buildings were in disrepair, but the relatively low real estate prices also attracted a younger group who were interested in owning and restoring the old houses. Soon, newspaper articles began to appear documenting the start-up of preservation activities in Monrovia, and in 1982 MOHPG held its first home tour to show what projects homeowners were undertaking. Soon after, the *Los Angeles Times* published an article about Monrovia's renaissance, and the Los Angeles Conservancy contacted MOHPG with an offer to co-sponsor a home tour in 1983. By 1986, the MOHPG tour had moved to Mother's Day where it has remained as an annual event for over thirty years (MOHPG 2015). With a grant from the state in 1984, roughly fifty MOHPG volunteers who acquired knowledge of architectural terms and concepts for describing historic resources for the first time, helped complete the first survey in 1985. Funds raised by the home tour paid for grants to homeowners undertaking renovations, and provided funds to donate to the historical society and buy books for the library.

In addition to continuing the annual home tours, MOHPG holds monthly board meetings open to the membership, publishes a monthly newsletter and provides a resource list to its membership. MOHPG organizes other events to promote historic preservation including tours of local landmarks and fieldtrips outside the city. MOHPG members also research the history of a member's own house for applications for landmark status and write and publish histories of homes for the home tour brochure. In 1998, MOHPG became a 501 (c)(3) nonprofit organization and in 2006 it changed its name to the Monrovia Historic Preservation Group but kept the older acronym. The change in the name, say some members, reflects the self-confidence of the growing respect and influence the organization has begun to have in the preservation of civic buildings and monuments.

Although the city council initially had been reluctant to consider historic preservation policies during the 1980s and 1990s, one of MOHPG's former presidents (1990–4) eventually served on the Planning Commission and was later elected to the city council. Because preservation sentiments and momentum were increasing in Monrovia, as a planning commissioner he was able to raise the possibility of at least studying the idea of an ordinance that would protect historic structures. In 1994, a Historic Ordinance Committee was formed with five MOHPG members and a city planner to study possible historic preservation laws and propose legislation for Monrovia. According to one member of that committee, the city council through city staff made its sentiments about protecting individual property rights clear. That is, even though the committee favored what they called "mandatory" protections for historic structures, the city council was only going to approve "voluntary" protections, protections the property owners themselves had to voluntarily request. Of course, homeowners can "voluntarily" apply to have their houses designated landmark structures, but there is no guarantee that the application will be approved. Rejections are often based on insufficient evidence of "significance" in local or regional history.

In addition, as noted in an earlier chapter, Monrovia's neighborhoods tend to be a patchwork of architectural styles built during different time periods. The mixture of styles seems to have made it difficult to visually identify or form historic districts even though the legislation provides for creating a district with only 50 percent of the structures contributing. The Monrovia Municipal Code on Designating Historic Landmarks and Historic Districts, Criteria 3 states about the structure:

> It contributes to the significance of an historic area, being a geographically definable area possessing a concentration of not less than 50% of historic or architecturally related grouping of properties which contribute to each other and are unified aesthetically by physical layout or development. (Monrovia Municipal Code, Section 17.40.060)

While the ordinance states that a little more than half of the properties must be "unified aesthetically," it implies that the remaining properties do not have to be. The reluctance of Monrovia residents to propose historic districts with almost half of the structures not contributing, or with less than half the property owners agreeing, seems to be based more on social concerns rather than historic preservation interests. In other cities, gerrymandered districts with an eclectic combination of styles are not infrequently pulled into the definition of a landmark district. Thus, it does not seem that Monrovia's ordinance prevents the formation of districts; rather it is residents' interpretation of the law conditioned by a perception that the city council will not approve the proposals.

Monrovia's Historic Preservation ordinance was adopted in 1995 and the seven-member Historic Preservation Commission (HPC) was created. In public hearings, the HPC reviews all proposals for landmark designations and certificates of appropriateness for remodels, and any proposals for the newly created Historic Commercial Downtown (HCD) zone. The HPC members are often drawn from among the MOHPG membership and include architects, realtors and other concerned residents. While the HPC used to meet every month, since the economic downturn it has moved to mostly quarterly meetings. The meetings are well attended by members of Monrovia's preservation community. Reviews of applications for landmark designations and certificates of appropriateness engage commissioners in extensive and "by-the-book" discussions of "significance" and "integrity" issues. For example, some applications for landmark designation do not rise to the high standards for local significance the commission sets and are turned down, thus dashing homeowners' hopes for receiving a Mills Act contract that depends on landmark status. Applicants can be critical of the HPC decisions as elitist because they feel the criteria are so strict.

Reviews of proposed remodels on designated landmarks for certificates of appropriateness typically give detailed consideration to the design and materials to be used, and focus on questions of potential damage, retention or enhancement of integrity of the historic structure. MOHPG members often attend HPC meetings and offer, or are called upon to offer, their own observations about the houses for which renovations are proposed. The owner of an 1887 Queen Anne Victorian proposed to build a deck made of steel, and Trex (a manufactured material) to wrap around the northwest corner of the house. As it does with all applications, the planning department reviewed the proposal, comparing it to legal requirements, and recommended approval as the deck did not adversely affect any significant features of the house. In the hearing, the HPC chair called on Monrovia's local historian who sat in the audience for his opinion. He said that the house "is the most original, highest profile landmark building in the city" and the deck is not an original feature. The speaker is considered by local preservationists to be the authority on every landmarked house in Monrovia.

The discussion that followed focused almost entirely on the material features of the proposed deck. The owner argued that originally there had been a porch on the north side of the house, the evidence of which, the "ghosts," can still be detected. Commissioners discussed whether the new deck would be seen from the street, what would the materials look like on an historic house, and whether the geometry of the deck design was appropriate for the style of the house. One of the commissioners wondered if the owner sold the house, would the new owner: "be happy? That is, will a future owner have to tear it all out?" The local historian commented that there were photographs available for reference. Additional criticisms raised doubts about the geometric shape of the design and the use of steel and Trex, which would never have been used in the original house. In the end, the HPC decided to postpone the vote until additional information and design considerations could be pursued.

Monrovia's HPC reviews usually follow closely the issues central to preservation cosmology as outlined in federal and state legislation and adapted to the local level. The composition of the HPC, however, is critical to this focus. At least one of the members of the HPC is considered a "purist" in terms of his views of preservation. As a restoration contractor, he tends to favor replacing with identical materials and forms those building elements destroyed by age and use. For instance, he argues it is better to replace broken windows with vintage glass rather than use new or tempered glass. Some of the other members of the commission, the realtor and a homeowner, tend to follow his lead, while other members such as the architect, tend to be more open to creative alterations that make the old house livable. These sentiments seem to mirror those found in MOHPG and the larger preservation community. One consideration all the commissioners tend to agree on is that if the proposed change does not alter what are thought to be the original features of a landmarked house, then at minimum the changes should not permanently change the house's character and integrity. It is likely that the commission's conservative strain in interpreting the historic preservation ordinance, coupled with the traditional resistance from the city council, constrains the issuance of certificates of appropriateness, landmark designations and the formation of districts that contain many periods and styles of housing.

As a small city, Monrovia's historic preservation community enjoys close and supportive working relationships among the city's planning staff, HPC and the members of MOHPG. Planners have maintained close advisory relations with the preservation community to try to resolve their burning issues, especially those concerning the apparent increased loss of historic buildings due to demolition and development. Preservation homeowners note that developer pressure is the top threat to Monrovia's historic fabric and they point to a number of efforts they have launched to try to combat it. One of the deepest concerns is the fear of "mansionization," the tendency of some developers or owners to demolish old houses and build newer ones that are out of scale with the neighboring houses. Monrovians identify a number of examples in their city and can name some of the builders responsible. One of them is a local Monrovia resident who has been building large-scale houses in Monrovia, and nearby cities of Arcadia and Bradbury. In fact, a number of preservationists pointed to Arcadia's hillsides west of Monrovia's borders to decry the number of mansions built there mostly by Chinese immigrants in the last two decades. In 2000, members of MOPHG rallied against a proposed development in Monrovia's near-pristine foothills and organized other Monrovia residents to oppose the project. They collected enough signatures to place on the ballot two bond issues enabling the city to acquire the land to prevent mansion development they saw next door. Their proposal included setting aside land for conservation and recreation uses. The bond issues passed with well over 70 percent of the vote and the city continued to acquire more land with funding from the state. In 2012, the Hillside Wilderness Preserve opened for recreation activities, while prohibiting all housing construction.

Although MOHPG members generally maintain a respectful presence at the HPC meetings, they have been known to vocalize their sentiments loudly when they detect an inappropriate project. In 2006, one of these concerned a proposed demolition of a 1913 Craftsman duplex and construction of a new Tudor-style house by an immigrant couple that had recently moved to Monrovia. The couple had applied for the demolition permit and was referred to the HPC because the house had been listed on the city's original 1984 survey as an historic resource. The duplex had been designed and built by the Tifal Brothers, a locally well-known design and construction company that built over 100 Craftsman bungalows in Monrovia. Because the HPC believed that the house met at least two criteria for designation as a landmark, they were able to place a hold on the demolition permit for 120 days. The planning department assigned the house status code of 5S1 "individual property that is listed or designated locally" (California State Office of Historic Preservation, Technical Assistance Bulletin 8: 5), which likely would have required an environmental impact report be conducted before the demolition permit could be issued. The owners did not appear at the first HPC meeting when their demolition permit application was denied, but did appear at the second meeting when they appealed the decision.

The June meeting brought out MOHPG members and the couple's neighbors in full force to advocate for preserving the house in its original condition. The owners were welcomed and given an opportunity to explain their rationale for seeking the demolition permit. First, they said they "loved being Monrovians" and "loved Monrovia." They cited the small size of each duplex unit being suitable for only a single person or couple, but not a family, not even if the two units were combined. They also cited the cost of remodeling, which they had determined was equal to if not more than the cost of demolition and new construction. The couple had hired an architect and proposed a new "historically appropriate" Tudor-styled house that they contended fit in well with the neighborhood in terms of scale and aesthetics. Their biggest complaint was that the house was infested with mice and fleas or tics that bit them constantly, and that despite working with exterminators, the problem had not been solved. They thought it could not be resolved without new construction that could seal the house. In proposing the new Tudor-styled house, they had clearly misread the sense of Monrovia's emphasis on maintaining its "historic" resources, which they took to mean a contemporary reproduction of an historic style rather than the appreciation of the original qualities of the house. At one point the owner declared, "The house is old, at the end of its life," which prompted others to give their own testimonials.

Members of the commission and then the audience began to speak, offer advice, and exhort the couple to preserve the house. Several commissioners suggested that the couple sell the house if they were not interested in preserving it. As one commissioner put it, "Have you considered selling the house, moving, rather than molesting an historic home?" Despite the couple's sad stories about the mice and tics, HPC members remained unmoved and unanimously committed to preventing the demolition. When the HPC opened up the discussion to the audience, no one there supported demolition either. One by one, neighbors and MOHPG members proclaimed their opposition. One neighbor said that movies are made in the neighborhood because of its historic character and a new house would not fit in. The MOHPG president offered advice and references to architects and contractors who could work to make the historic home functional. He also said he could provide support for landmark designation that would make it possible to obtain a Mills Act contract. Two other speakers reported that they also had problems with mice and termites, but hired exterminators who solved the problem. Finally, echoing sentiments of the others, another neighbor suggested, "You can have every modern convenience in an older home; there is no reason to have less than you deserve. But if you do not want to preserve the house, we would like you to leave. You could come back, of course, if you want to preserve historic houses."

The intense emotions and passionate arguments on the side of preservation seemed to have convinced the couple to withdraw their proposal. The owner said that they had decided to work on remodeling the house, adding, "We do not want to fight the commission. It is not our way." Within months, however, the owners put the house up for sale and one of their neighbors bought it, applied for landmark status and received a Mills Act contract to help defray the costs of renovation. The conflict highlights the very different points of view between preservationists and ordinary homeowners who may not value the "original" qualities of an aging and sometimes deteriorated house. When the couple could not acknowledge the aesthetic potential of their older home, to preservationist Monrovians they failed to exhibit the proper sense of responsibility and stewardship for a community resource. This lack of sensitivity to local "history" seems to have provoked intense exclusionary sentiments and the pursuit of legal sanctions to protect the material representation of an earlier time.

MOHPG promotes this case as an example of how existing laws can be used successfully to save endangered properties, however, as many preservationists point out, it only works on properties deemed historically "significant." What should happen to the rest of the neighborhood fabric has not yet been resolved. One city councilman said that Monrovia needs at least one more councilman in favor of historic preservation to make it possible to enact "mandatory" protections as MOHPG would like. This example is also an instance of how preservation enthusiasts use laws and community sentiments to exclude aesthetic expressions of people whose preferences are at odds with the local community. At work, sometimes, is a kind of socially constructed emotional extortion that tries to shame people into doing "the right thing" even if they do not believe in it. This may be the nature of small town Monrovia, but it works because the public discussion of historic preservation in the HPC tends to be coercive as much as it is persuasive, by invoking social norms as in any small group.

City of Pasadena

The city of Pasadena adopted its first ordinance protecting historic structures in 1969, and revised it in 2005. The city has one citywide historic preservation organization, Pasadena Heritage, which was founded in 1977 by a group of residents who became alarmed at the disappearance of some of the city's richest architectural resources. The organization's founder describes moving to Pasadena in 1971 because of the affordable real estate, although the city's reputation was not very positive at the time. Pasadena's history included being an early twentieth century winter resort for Midwesterners, the home of Cal Tech, and host to the annual Rose Parade and Rose Bowl Game, but since the sixties the city had lost its allure. The local economy had been declining, the old urban core was losing businesses to shopping malls, and older neighborhoods had seen better times. Even though Pasadena had had a celebrated history associated with wealthy elite families, such as the Gambles and Wrigleys, and important cultural institutions, by the 1970s many of those families had fled to Orange County to avoid court-ordered mandated busing to integrate the public schools. Many mansions and estates belonging to early twentieth-century elite families were sold and demolished, and in their places large modern condominium complexes and large-scale commercial developments were built. To fight some of the demolitions, residents of the West Pasadena Residents Association came together with other neighborhood groups to oppose local condominium development.

Many Pasadena residents in those same neighborhood groups had also become very upset over a large-scale shopping mall project that Pasadena's Redevelopment Agency had worked with the city council to advance. The mid-1970s' project required the demolition of several historic commercial buildings, including the Pasadena Athletic Club, along Colorado Boulevard in order to construct a totally enclosed structure that was intended to revitalize the declining commercial area. Residents opposed to the project engaged in a bitter battle that ultimately unified the preservationists. Although they lost the battle over the shopping mall, the conflict set the stage for the formation of a citywide organization, Pasadena Heritage in 1977. As a result of their efforts, the city council moved to undo the power of the redevelopment agency and give more weight to the input it now sought from the neighborhoods. The very neighborhood associations that provided the foundation for organizing the opposition to development numbered only nineteen in the 1970s, but grew to well over eighty associations by 2006.

In the early years, Pasadena Heritage identified education as their primary mission. Members conducted research about and led walking tours in residential neighborhoods, some of which were suffering from deterioration due to bank redlining practices used against minorities and lower income families who had moved in. They talked to merchants in old Pasadena, held classes in restoration techniques for residents and assembled lists of craftsmen and contractors who specialized in restoration. The Pasadena Heritage board began advocating in the city on other projects, the Castle Green, the redevelopment of Old Pasadena and the Civic Center. To raise awareness, every year the organization still hosts the Craftsman Weekend event that includes lectures, receptions, workshops and tours of historic neighborhoods and some residential interiors. Rather than organize their own home tours, some of Pasadena's landmark neighborhoods now participate in the Craftsman Weekend event. In 1998 the organization started the Heritage Housing Partners that builds affordable new properties or rehabilitates historic ones.

Pasadena Heritage's early education and preservation activities roughly coincide with the 1983 arrival of one of the preservationists credited with the creation of the first historic residential district in the city in 1989. He had moved from Riverside after living for a while in the Peter J. Weber house, the current headquarters of the Old Riverside Foundation. He tells the story of how shortly after he bought his first house a developer demolished a large Craftsman house on the corner of a main thoroughfare in order to build a large apartment house.[1] Some of his neighbors were also outraged and they soon discovered that the new zoning on the nearby thoroughfare allowed for many more apartment buildings to be built. They organized themselves and petitioned the city to see if the street could be "downzoned" to permit only single family housing again. They won the argument with the city council, which reestablished the zone permitting only three residential units per acre, thus making the construction of apartment buildings unworkable.

The collaborative work on downzoning brought together a number of neighbors who shared deep concerns about the increasing destruction and possibility of preserving the historic features of older houses in the neighborhood. In particular, residents noticed that some of their neighbors were busy applying stucco to the exterior shingles and wood siding of their bungalows, destroying the historically significant details in the process. Coincidentally, the city had just begun to survey the same bungalow neighborhood where the activists lived, which was also home to a city councilman and a city historic preservation staff person. The possibility of establishing the city's first historic landmark district was discussed, and the city provided support in developing a conservation plan. A group of neighborhood volunteers set out to collect signatures from at least 51 percent of the 962 property owners, as mandated by the preservation ordinance, in order to support the proposed historic district, but they had no idea how many residents would actually sign. When they canvassed the houses, the volunteers heard many other concerns about rising crime, traffic, and the condition of the small park at the center of the district where gangs and drug activity were found. They realized they would need a neighborhood association to address the larger issues, and the city made forming such an association a condition for granting landmark status to the district. Volunteers also found many residents who clearly resisted the idea of local government telling them what they could and could not do with their houses, but surprisingly the volunteers were able to collect signatures from 55 percent of the neighbors. In 1989, Pasadena established its first Landmark District in what was affectionately called, Cottage Heights. Since that time, the boundaries of the district have been expanded to include 1,100 properties, and the neighborhood association currently has eighty households as current members.

The Cottage Heights Neighborhood Association (CHNA) is the official representative of the landmark district and coordinates all of its educational and advocacy activities. Its charter and by-laws were revised in 2005 and it states as one of its primary purposes the preservation of the "architectural integrity of the neighborhood." CHNA is incorporated as a 501 (c)(3) organization. The association's fourteen-member board representing all "sub-areas" of the neighborhood meets every month and holds four general membership meetings each year. The district is divided into north and south subareas with each street having at least two block captains who are charged with communicating with residents and reporting issues back to the board. Generally, a core group of about 25–30 members is the most active in the organization, which often leads to burnout, but these members tend to be the most passionate and ideologically committed to the preservation cosmology. Indeed, many of these members seem to have maintained long-term influence in the organization.

The neighborhood association has hosted its own annual home tour twenty-four times since 1989, and attracts people from all over southern California and beyond, vying with Monrovia for the most popular local home tour. The neighborhood's own

Figure 17 Home tour in Cottage Heights, Pasadena. Credit: Denise Lawrence-Zúñiga.

"historian" had been one of the early volunteers to survey properties and collect signatures and often writes some of the rich historical descriptions for the houses shown on the tours. The tours tend to be elaborate events with workshops and lectures, re-enactments of historical events in situ and cookie giveaways. One former president says the tour is the glue that holds the association together. It is the main recurring activity that requires members and nonmember neighbors to cooperate, and it is a source of great pride and gratification. The association uses the tour as its major annual moneymaking activity and has used some of that money in the past to pay for renovations in the park at the center of the district. The park is the site for Halloween activities, Snow Day, and Sunday in the Park for district families, and the association holds an annual Founders Day picnic there at the end of September.

Monthly meetings are occasions for discussing neighborhood problems and making preparations for special events. The most frequent discussions, however, seem to center on concerns for inappropriate remodeling in the neighborhood: Who is replacing original wood windows with vinyl ones? Who is adding a second floor or putting in an illegal fence in their front yard? Who is demolishing their house? Some of these discussions, according to a former president, pit the purists who favor living with the exact material conditions of history against those who think changing the material features of the house to accommodate a contemporary lifestyle is appropriate. One board member characterized these distinctions as between those who view the house as a living organism—that changes as lifestyles change—as opposed to seeing the house as museum for a fantasized past. A member's view of the materiality of their own house is often projected onto neighbors' houses and their occupants. Their concerns may be provoked when neighbors who have lived in the area a long time remodel or modernize their old houses. They may also have a sense of ownership and belonging in the neighborhood, of privilege, that the recently arrived do not have, but that ownership does not necessarily translate into appropriate restoration practices. Purists may be offended that a neighbor has used Home Depot materials or techniques to repair or renovate their house. Sometimes the original owners have died and their heirs have become landlords, but are no longer invested in the neighborhood. Some residents, then, are renters, while other newer owners may be immigrants with very different aesthetic ideas and lifestyles. The diversity of homeowners and occupants in the historic district presents challenges to association members and preservation oriented homeowners.

There are no clear strategies for association members who at once want to understand and reach out to their neighbors, but also want them to conform to the aesthetic ideals they hold dear. Some members see a solution in the gradual gentrification of the neighborhood; as new residents buy in and property values rise, the educational levels and occupations will begin to correlate with the economic value the houses represent. Others who want to exert more effort in educating all their neighbors contest the assertion that class homogeneity will solve the problem. They reach out by distributing flyers publicizing meetings and workshops, and they talk to neighbors to convert them to preservation values. Individual homeowners report that they are sometimes able to persuade their immediate neighbors with whom they have developed a personal relationship to preserve their homes. Real estate investment has also been gradually making a difference as long-term homeowners gain equity in their houses and decide to cash-in. One resident said that several houses on his block had been bought and "flipped" with a minimal amount of investment before being sold. Even though the newly remodeled houses were not restorations, the results produced much better quality housing and attracted wealthier homeowners. Some association members, however, lament the permanent loss of character-defining details in the remodeling process. Continuing complaints focus on renters, or landlords, whose lack of maintenance many consider destructive to the older housing stock. If landlords live in the neighborhood, they may be approached, but often they are absent or do not care.

The most promising situations involve homeowners engaged in remodeling who may be ignorant or misinformed about appropriate restoration practices. Several board members related a story about the proposed addition of a second story on a colonial bungalow remodel in which the architect had been told by the city that, based on the Secretary of the Interior's Standards, a material distinction had to be made between the new and original construction. Association members said they were very concerned about the height of the second story addition and "arrangement of windows" in the design that seemed to result from the city's advice. Once association members met with the architect they informed him that they thought he had been misled by the city and did not need to make such a dramatic distinction between the new and original, which resulted in a simpler, and for them more satisfying, design solution. Another common concern is "mansionization" or increasing the square footage of small houses by adding on a second floor, that some say, "overwhelms the original part of the house." There is often a moral condemnation couched in this criticism implying that people just want to consume more and more space, and a suggestion that if they need more space, perhaps they should just move. Others are more accepting of the need to expand the smallish bungalows, especially for families with children. They hope, as in the previous case, that owners will do it sensitively.

Another theme that association members discuss is that they, like preservationists in other cities, believe the city should be responsible for policing and enforcing the historic preservation requirements in the landmark district. A number of homeowners complain that city employees are not always well informed about the provisions of the historic preservation ordinance. As one resident claimed, "You can't just call the city and ask them a question, you get five different answers." Building department officials are the most notoriously misinformed about what exemptions are permitted to owners of historic houses. For instance, if the original garage only housed one car, according to the California Historical Building Code (2013), its replacement or remodel can be the same size even though contemporary building standards require at minimum a two-car garage. Some association members ask what the purpose is of living in a landmarked district, if people can do what they want to their houses? "Why should I follow the rules, but others don't have to?" And they expect the city's code enforcement to patrol the neighborhood to prevent people from building inappropriate remodels, or they call in to report (in secret) on their neighbors. In fact, of all the city's historic districts Cottage Heights residents call city hall the most often to complain, followed by four other landmarked districts in the immediate vicinity. These neighborhoods tend to be ethnically diverse and are home to long-term and younger families of different classes, many who may be completely ignorant of the fact that they live in a landmarked district that has special restrictions on remodeling houses.

Association members often complain that residents do a lot of remodeling on the weekends; one estimated that 90 percent of all construction occurred on the weekend and much of it without a permit. Members had to wait until Monday when city offices were open to call in complaints, but by that time it was often too late. One crisis in Cottage Heights that people still discuss occurred over one weekend when the owner of a somewhat deteriorated Craftsman began to demolish the interior of the house. Although he had a permit, by the time the demolition was completed on Sunday, only the front wall remained. The neighbors became enraged as they watched the demolition unfold, and called the city over the weekend trying to locate someone to come out and issue a stop-work order before the house was completely destroyed. The owner then proposed to make the remodeled house even bigger than originally proposed because so much of the house had had to be demolished due to termite damage. A newly revised ordinance specified that the punishment for exceeding the scope of permitted demolition would be to withhold a building permit for the property for five years, but the neighbors did not want to have to live next to a vacant lot. In a way, the city had failed to enforce the ordinance and the neighborhood was left with a mess. As a result, some association members put forth an effort to negotiate with the city and the owner to find a workable design solution for the property. Since then, the city now makes available a staff member from code enforcement on-call over the weekend.

One reason residents give for joining the neighborhood association is the concern for "code violations" and general neglect in their neighbors' houses. They look to the association to enforce their vision of the neighborhood and promote it by teaching the aesthetic values necessary to reform their neighbors' taste. There are, however, some significant differences in wealth throughout the historic district, and therefore inclination and capacity to embrace preservation values. In the district's northeast section households tend to be affluent, and many have strong preservation values, while lower income families, with more renters and ethnic diversity, live in the south and southwest sections. In fact, some association members living in the southwest section try to recruit their neighbors but are frequently rebuffed. Even so, wealthier homeowners may not join the association because they object to the board's insensitivity to meddling in homeowners' property rights, while others say the association is not doing enough to prevent inappropriate remodels.

Some households who would otherwise like to participate in the neighborhood association feel excluded, and a number of people have said that the association is elitist or their values are too purist. Many have joined Pasadena Heritage instead. One board member claimed, "In all honesty, I sometimes think people on the board can be a little too militant and they put people off." And there are a significant number of residents who are sympathetic to preservation but refuse to be involved with any organization. There are also families who have lived in their houses thirty or more years who have never been interested in historic preservation, and do not belong to the association, but their houses have been included on the home tour. As one member put it, "they just love their houses." Thus, the landmarked historic district is far from homogeneous as far as preservation values are concerned.

Another reason for joining the neighborhood association has been to meet people with similar values and interests, which serve to strengthen the vision and its moral underpinnings. Several residents revealed that they never before felt like they were part of a community until they moved into the neighborhood. Part of that sense of belonging comes from working on their houses and gradually meeting all the neighbors, some long-term and other newer residents, who stop by to see what they are doing. Long-term residents can tell stories about the previous owners and share bits of gossip that make restorations even more meaningful. But in a larger sense, these homeowners simply wanted to be part of an historic preservation neighborhood. One homeowner said, "We just wanted to be involved, to know what was going on, and wanted to help out—we wanted to support it." Another member said she just wanted to meet people and be involved. She had moved to Cottage Heights as a single woman living alone and it was important to her to know people around her, in the neighborhood. But she also said that after she attended a few meetings she wanted to "break into the clique and meet the cool kids." She volunteered to be a docent for the home tour and then became the docent coordinator the next year, which was followed by being tapped to direct the home tour and, finally, she was elected secretary of the association.

Clearly, Cottage Heights' neighborhood association brings together residents who are the most committed to historic preservation and pursue much more consciously the cosmology for restoring the original conditions of early twentieth-century houses. To them, the policing and reforming neighbors' building practices, the educational tours and workshops, are all aimed at encouraging, nudging, coercing and persuading homeowners unlike themselves to change their ways to achieve an idealized community. The vision that they hold of the neighborhood may be more fantasy than reality, but it is at least partially based on understanding the legal under-pinnings of the historic district and the significance and integrity of the houses that make up the whole neighborhood. One resident, Paula, claimed:

> Cottage Heights has been described as one of the best examples and one of those intact examples of turn-of-the-century middle-class housing in the country. The idea that if you wanted to know how the middle class lived—truly the middle class—you know, just how your basic tailors lived at the beginning of the last century, you can come here and see what it looked like.

Looking at the neighborhood as a "collection" of houses, rather than in terms of individual houses, constructs a unified vision that preservation homeowners promote as a source of pride. The view of a fully restored early twentieth-century residential landscape is a critical desire and expectation of Cottage Heights' homeowners, even if the purity of the vision varies. "I think that every house in the neighborhood in that collection is important because as we lose them, if we lose them, and we are losing them because of the city, that fabric erodes." They emphasize the material features of home and neighborhood, rather than their social or cultural aspects, but they often impute certain social meanings to physical elements. Sometimes specific design features of neighborhood houses are identified as particularly important symbols of community, especially houses with highly desirable front porches. Paula, who had bought a second house on a northwest street in Cottage Heights elaborated:

> What I'm drawn to now, the new house, is the promise of the neighborhood, big trees, and the fact that George and Anthony … say they just invite themselves onto people's porches and sit there and drink coffee. And I can't wait—I cannot wait for that. I cannot wait to have friends right up the street that I can walk to … or if I wanted to have a block party, because we do that in CH, people will come.

Thus, the construction of the restored historic landscape of the landmark district is both goal and fantasy, the material qualities of which correspond to or activate a set of neighborly social behaviors that residents say they seek and cherish.

The unified conceptual construction of an historic neighborhood is, however, contradicted by the multiethnic, multiclass realities. Preservation homeowners are well aware of the contradiction and, while they express a certain sense of satisfaction about living in a diverse neighborhood, they are also consumed by the challenges it presents in achieving architectural integrity and coherence. Several association members mentioned how well they knew their neighbors from different ethnic groups. Laura observes:

> That's the one thing we like about this neighborhood, whether you like your

neighbors or not, you do know your neighbors. It's very diverse, very ethnically diverse. We have Armenians, Hispanic, Black, Filipino, Thai, Japanese.

Some of the class and ethnic differences have been linked to a history of gang activity and drugs, problems that especially plagued the area during the 1970s, but residents argue that these have significantly subsided since the district was landmarked. In fact, some members say that revitalizing the park helped make it safe and got rid of a lot of the drug dealing. Nevertheless, association members who live in the northern section are less likely to recognize the conflicts in lifestyle, or say the problems exist in "pockets."

Association members can readily identify the material evidence of different aesthetic preferences associated with particular neighborhood groups, such as the propensity of Hispanics to fence in or build a brick wall topped with wrought iron fencing around their yards and to use the space as a living room. Marie said:

> If there are cultural tensions it's because certain cultures live differently. Like I said, Hispanics live in their front yards. Their front yard is an extension of their house … They have showers—baby showers in the driveway and a bounce house in the front yard for no good reason except that it is Sunday and they are constantly out there and, you know, the kids are running around and that's the way they live and yet, that doesn't happen anywhere in the neighborhood. If there are cultural tensions, or ethnic tensions, it's because the lifestyles clash.

The association's solution to the problem is always education, letting "people know about where they live and why it's historically significant, and how important it is that they contribute to that by taking care of their homes." But the conflict raises deep philosophical issues about which many members express ambivalence or may not be completely aware. Preservation homeowners, while proclaiming unity about historic preservation values, make efforts to avoid showing a bias in condemning their non-conforming neighbors, and strive to signal inclusiveness about them. The kinds of complaints about the neighbors' lifestyles, as described above, indicate a not unconscious preference for more middle-class behaviors and values, as well as design features.

Some homeowners claim private property rights entitle them to remodel their houses as they see fit which contrasts with the preservation idea that each house represents an important piece contributing to the visual coherence of the neighborhood. The cultural construction of a unified historic district entitled to its own preservation implies a kind of material agency, and a moral project, imposing on homeowners the rights and duties to ensure the continuity of the neighborhood as an historic whole. As one resident argued, when people buy into Cottage Heights they should realize they are buying into a piece of Pasadena's history and that it comes with a set of obligations and responsibilities. She further opined: "I want you to buy the neighborhood as much as you are buying your house. And by buying the neighborhood, I want you to be buying into what this is. You are buying into a historic district." The essential qualities of the historic district are for preservationists self-evident, "I think we really need to be led by what this neighborhood is at its essence. We saw something worth saving in 1989 ... and that is what we need to continue to look at." In this conception of the historic community, homeowners become like mutual shareholders in the district to which they individually contribute their own appropriately restored house, and benefit collectively from the historic context provided by the ensemble of all the neighborhood's houses.

Cottage Heights, as an historic district, is one of the most complete and well-known productions in southern California. The city of Pasadena nominated the district to be placed on the National Register in 2008. Well before that, media coverage of the neighborhood in a wide variety of magazines and journals made the historic district widely known respected and admired for the ostensible historic "community" it has built. Magazine publications have selected it as among top U.S. neighborhoods, celebrating its historic architectural styles and sidewalks that people actually use to take walks. In 2009 the American Planning Association considered Cottage Heights one of the top new communities in America. The recognition of the district from outside its boundaries enters into residents' discourse and gives the place a special standing, a status, and enhances a sense of entitlement. This recognition also adds to residents' sense of pride in the protection of historical materiality that bolsters exclusionary feelings about the place, and its residents, as exceptional, and the desire that gentrifying residents will help complete the historic imaginary.

Conclusion

The residents of both Monrovia and Pasadena's Cottage Heights express strong historic preservation values in the rehabilitation of their private residences and extend those sentiments into the civic sphere to define their community's identity, but also to collectively bolster their moral project. Their sentiments, however, do not occur in a vacuum, but arise in facing the challenges of recapturing or salvaging the material remains of homes from the ravages of time and injuries suffered at the hands of well-meaning, and not so well-meaning, prior occupants. In both cities variations in the knowledge and understanding of historic preservation cosmology and legislative protections inform public discourse. While professionals focus more precisely on material conditions and architectural details that are explicitly regulated in legislation and official guidelines, homeowners tend to be more "flexible" in their interpretations. In short, they know what they don't like, but are less skilled in finding the appropriate language to describe it. And, although homeowners in Riverside and Ontario tend to be even more vague in citing particular design issues than the residents of Monrovia and Pasadena, they are all equally aware that their aesthetic objections can be subjected to political pressure and interpretation by commissioners and city council members.

The gentry freely interpret the meaning of material conservation in public settings to achieve more than the preservation of the historic landscape. Bound by preservation sentiments and enthusiasm, the newcomers have organized themselves to take an active interest in neighborhood and citywide affairs. But their organization implicitly promotes exclusionary sentiments about alternative aesthetic expressions, suggesting a desire and belief that gentrification, which results in replacing residents, or converting current neighbors to a preservation ethos, may be the only way to achieve their ideals. Without necessarily intending their sentiments to push out and displace non-believing neighbors, the effect of homeowners' passionate beliefs in the virtuous service of preserving the past produces similar results. The inevitable but unintended consequence of historic preservation may be neighborhoods that have become homogeneously middle-class and symbolically "white" once more, despite the actual ethnic composition of their residents.

Immigrant Challenges: Communicating Preservation Values across the Cultural Divide in Alhambra[1]

Although many southern California cities with historic architectural resources have by now adopted preservation legislation, not all have committed to it. Some cities resist legal protections because city officials lack the political will in the face of local developer opposition, or they are reluctant to intervene or obstruct the private property rights of owners. The cities of Ontario and Monrovia had to overcome these challenges in passing preservation legislation, and Riverside continues to encounter resistance as regulations are put into action. In some cities, officials and residents do not see any threat and trust that property owners will be mindful of a city's own building traditions, or they may even welcome new construction as a means of "revitalizing" older building stock or historic zones. These cities generally find their zoning codes already provide enough guidance for construction in terms of densities, lot coverage, setbacks, and building mass. In other cities, however, preservation advocacy has made it necessary to pursue a middle way by implementing residential design guidelines, rather than preservation laws, that can act as a reference text to suggest and recommend appropriate design options.

Alhambra is one of these places where increasing local advocacy challenges political resistance to valuing historic architecture. Initially, resistance to historic preservation began in 1984 when the city conducted its first survey of historic resources, but the city council refused to accept the findings. Local residents who remember those days claim that the city council members were already too involved in real estate investment and development to endorse any limitations on construction activity. The rejection of the survey signaled that the city had no interest in protecting historic properties, but it also deprived the city of the official database of properties that might require legal protection under state laws such as CEQA. In the years that followed, much of Alhambra's downtown historic corridor was demolished and built anew, and older residential areas were converted to rental housing, construction eagerly encouraged by the developer-friendly city council, which also acted as the city's official Redevelopment Agency.

Concerned about the rapid transformation of older historic neighborhoods, residents formed a preservation organization in 2003, the Alhambra Preservation

Group (APG), to promote an understanding of the city's history and its architecture, and to advocate for municipal protections. Like Riverside and Ontario, the preservation group was first housed in Alhambra's Historical Society, which had its own museum building and archives of city history. This convenient arrangement allowed the preservationists to take advantage of the Historical Society's non-profit status. It soon became clear, however, that the preservationists had a very different vision and set of values than the members of the Historical Society who tended to be older and did not see architecture as an historic resource. In fact, many of the Historical Society's members were interconnected with the original city council members and their colleagues who promoted urban revitalization and development.

Much of the preservation group's advocacy took place at city hall in the form of attending Design Review Board meetings that allowed their members to speak up for the residential properties owners were seeking to remodel. They also organized home tours, special lectures and events to promote Alhambra history, and have provided a candidate's forum for city council elections in order to raise awareness. In 2006, after three years of aggressive advocacy, the city council authorized the adoption of residential design guidelines the preparation of which were contracted to an outside professional design firm. The company that prepared the guidelines consulted with city officials, members of the Design Review Board, and members of APG, the advocacy group. The Single Family Residential Design Guidelines (RDG), published in 2009, carefully follow the broad principles of historic preservation legislation by establishing neighborhoods or districts exhibiting design consistency and recommending "appropriate" stylistic features, although following them is not legally required. That is, a homeowner cannot be legally compelled to adopt specific design features as a condition for winning design review approval, even though the features or alternatives are strongly recommended.

Unlike the other cities considered in this study, Alhambra's residential neighborhoods have attracted renewed interest in preservation protections largely because of the construction and remodeling activities of recently arrived immigrant groups. These immigrants have significant financial resources and frequently threaten local traditional aesthetic norms with massively built new mansions or other kinds of modernizing remodels. Currently, the primary threat comes from Chinese immigrants, whose wealth and real estate investment strategies are aimed to appeal to co-ethnics living locally and abroad. Alhambra has long been home to whites and Latinos, but the incursion of Chinese, initially from Monterey Park on the southern boundary of the city in the 1970s, has intensified as Asian immigration broadened. The earliest Chinese immigrants came from Taiwan, and settled in Monterey Park, but since the 1990s immigrants from Hong Kong and Mainland China have increased and have settled into the larger San Gabriel Valley area. Wei Li (2009) calls the area an "ethnoburb," a multi-ethnic community where no one ethnic group holds the majority position. In recent years, however, the Asian population has achieved majority status in a number of communities such as neighboring San Marino, Arcadia, and Rosemead. Chinese immigrants have been speculating in real estate, buying and remodeling houses throughout a wide area of about thirty cities, and then advertising homes for sale locally and overseas in China. In many cities the remodels have replaced older bungalows or postwar tract houses with very large-scale mansions, designed to maximize size and scale as permitted by zoning regulations. Chinese property owners argue that co-ethnic transnationals are eager to acquire as much square footage, and opulence, as they can afford, given the constraints on acquiring these in China. Moreover, they argue that Chinese families are large, so mansions can be justified for housing extended families. Alhambra's long-term residents who tend to be largely white and middle class mostly oppose these remodels.

In ways similar to historic preservation regulations, residential design guidelines and review are a form of what Ghertner (2010) calls "aesthetic governmentality"; procedures that seek to educate and regulate the production of architectural forms, and encourage conformity to the community's aesthetic norms. Ghertner develops his ideas in a study of slum clearance in New Delhi, where government officials try to convince residents to accept replacement housing elsewhere so as to make the land they live on available for development. He links these moves to the neoliberal turn in local governance that emphasizes privatization and market principles to achieve public goals. Aesthetic governmentality relies on encouraging individual responsibility and choice. Neoliberal proponents argue that individuals should be governed as members of communities who exercise freedom of choice in the marketplace to satisfy "aspirations to self-actualization" (Rose 1996: 147). Design guidelines selectively constrain those choices and impose a dominant aesthetic order. They reconstruct the homeowner's sense of self and place, while advancing a city's vision that ostensibly serves the common good (Ghertner 2010: 207). In describing Alhambra's existing housing as desirable and idealized models, design guidelines seek to establish aesthetic norms, legitimize these styles, and align homeowners' perceptions of the appropriate suburban environment and their place in them.

Aesthetic governmentality employs a series of "techniques" and "devices" that aid in the regulation of individuals' decisions and practices. Visual imagery codified in photographs and drawings projects aesthetic standards that represent an idealized construct of an early twentieth-century southern California suburban community. These images portray experiential truths rather than rational ones and, while

Figure 18 A new mansion surrounded by single-story bungalows. Credit: Denise Lawrence-Zúñiga.

accompanied by narrative information, they are meant to draw the viewer into a photograph and facilitate self-identification. The guidelines also employ micro-practices to focus attention on specific design details that work on homeowners' inter-subjective experiential knowledge of spaces, in order to change the way they see space and themselves in that space, and to convert them to a new set of aesthetic values. But, aesthetic governmentality also requires a ritual act, a performance such as a public meeting or hearing, where the homeowner's design proposal is compared to the idealized guidelines, the legally established aesthetic norms, so that the aesthetic shortcomings of the design proposal can be revealed. These ritualized proceedings have the effect of training applicants to see themselves as part of the problem while encouraging them to participate in producing the solution (Ghertner 2010: 204). Moreover, where guidelines consist of recommended design solutions, rather than requirements, the ritualized hearing provides a social setting where compliance is encouraged through a subtle but coercive form of social control.

Immigrant homeowners with distinct cultural backgrounds may have a variety of expectations about municipal regulatory processes. From their own experiences in their home country, they may have had dealings with building and planning departments and officials, but rarely are specific design features of a building required or recommended. Moreover, the expectations immigrants have for their houses tend to be strongly embedded in both the cultural norms of the home country and their own particular aspirations and preferences that the new locale inspires (cf. Zhang 2010). To conform to a new set of unfamiliar aesthetic requirements through novel procedures is often a struggle. The struggle may be related to language or custom differences, or it may be the result in a conflict of values, especially those related to private property. These requirements act as tests of cultural competence, and that conforming to design guidelines and passing through design review work on the applicant's "self-making subjectivities" (Ong 1996: 737). All applicants are required to submit to these processes to secure permits, and so the power relations embedded in the regulations tend to induce conformance with aesthetic norms. The obvious challenge for immigrants, however, is their lack of familiarity with the process or knowledge about historic architectural styles, which are deemed cultural products and constitutive of local heritage.

Aesthetic governmentality is particularly effective when the design review performance links the homeowner's attention to the aesthetic shortcomings in their own design by comparing them to aesthetic norms codified and legitimized in the published guidelines (Ghertner 2010: 204). Like historic preservation reviews, design review engages applicants with design experts who explore reasons for an applicant's decisions and recommend more acceptable alternatives, if needed, before approving the proposal. Design review hearings resemble a theater at which the applicant appears and is subjected to advice dispensed by experts whose superior knowledge is meant to be respected, while the audience acts as a representation of the "community" who ostensibly embody and express the aesthetic norms advanced by the design guidelines. These processes are necessary in order to help the applicant "see" space and themselves in that space to rework and change the applicant's inter-subjective experiential knowledge of residential aesthetics (Lawrence-Zúñiga 2014, 2015). Hence, residential design guidelines champion their educational reform value by encouraging residents to ostensibly learn and incorporate aesthetic norms as they restore and build houses and neighborhoods.

Alhambra preservation group

Alhambra's preservation advocates include younger newly arrived homeowners, and older long-term residents, who are middle or upper-middle class, and mostly white. Although advocates own older homes, they are not necessarily clustered in one part of the city but scattered throughout. By 1903, Alhambra was incorporated and land, initially dedicated to citrus farming, was subdivided and sold to build housing tracts. Houses of all sizes and styles were constructed in the early twentieth-century, and, today, these constitute the primary affordable resources for preservation strategies, and also for demolition and reconstruction. Initially, settled by white settlers from the Midwest, Alhambra welcomed large numbers of Italian–Americans in the 1950s, and Mexican–Americans in the 1960s. While significant numbers of new postwar housing tracts were added, both of these groups also engaged in substantial remodeling and modernizing of older houses. It was not until the arrival of large numbers of Taiwanese in the 1970s that some of the most deteriorated housing stock was converted into rental units to accommodate increased flows of Asian immigrants. Initially the city welcomed this "redevelopment" activity as a way to modernize and spruce up its image, but civic leaders also became interested in the infusion of Chinese money. By the 1990s, wealthy immigrants from Hong Kong and Mainland China, and lower income Asians from throughout Southeast Asia settled in the area (Li 2009). A commerce-friendly attitude facilitated the establishment of banks, insurance companies and other financial institutions so that Alhambra, today, is well known throughout the Chinese community as the ethnoburb's finance center.

By the 1990s, Chinese businesses had transformed commercial streetscapes with Chinese language signage, but long-term residents were more concerned about increasing densities and building styles produced by the new Chinese residents in some neighborhoods. They objected to the subdivision of single family houses into apartment buildings, the construction of one- and two-story rental units in the back yards of older houses, and the demolition or extreme remodeling of small bungalows to vastly increase their sizes and change architectural styles. In Vancouver and Northern California cities where Chinese and South Asians have also settled and built similar mansions opposed by long-term residents, the style ascribed to these largely stucco blocks is called "Asian–Mediterranean" or faux Mediterranean (Smart and Smart 1996; Mitchell 1993; Ong 1996; Chung 2011; Lung-Amam 2013). In Alhambra, preservationists perceived these styles as largely incompatible with traditional Craftsman Bungalow, Tudor, or Mediterranean Revival styles that predominate, especially when the homeowners seem to have obliterated all the original historic features of the house.

Preservationists have been offended by new immigrants' apparent rejection of local architectural traditions, they are also put off by their design decisions, speculative activities and transitory presence in many neighborhoods. Mobilizing the APG they have organized events to educate their new neighbors, and older ones, about the economic and cultural value of restoration and preservation of older bungalows by hosting home tours and holding community events. More importantly, however, they have committed their attentions to the monthly Design Review Board (DRB) meetings where homeowners, or their agents—often a designer or contractor— appear as applicants seeking construction permits. As a result of APG members' regular presence at these meetings, the DRB has begun to use more stringent design principles and exacting standards, many of which are outlined in the guidelines, to bolster their reviews and critiques of homeowner proposals. In 2010, the five-member board was comprised of four Asian design or development professionals, the fifth member is a white, long-term resident of the city; the city's staff architect, who is Japanese–American, reviews and comments on design proposals in advance of DRB meetings.

Design guidelines and review

It is worthwhile examining how the techniques of aesthetic governmentality operate in a city without historic preservation traditions and where new immigrants are affluent enough to contest aesthetic norms. Alhambra's Residential Design Guidelines have as their purpose the description of an aesthetic "framework of community design concepts to site planning, landscape architecture and architecture in the R–1 (Single-Family Residential) zoned neighborhoods within the City of Alhambra" (Single Family Residential Design Guidelines 2009: 1–1). The application of the framework for assessing design proposals is a means to produce high standards that improve the city's character while keeping new construction consistent with the traditional qualities and unique features of the neighborhoods. The guidelines outline preferred and undesirable features including building oversized second-floor additions, employing architectural styles that deviate from the original, and using inappropriate building materials. While zoning regulations specifying lot setbacks, number of units permitted, and massing limits are mandatory requirements for design proposals, the aesthetic features are only recommended and give, in principle, owners and DRB members the possibility for negotiation.

Like historic preservation ordinances, RDG include written descriptions as well as photographs and drawings to illustrate recommended design ideas. Twenty-six Alhambra neighborhoods are identified with maps, a brief historical account, archi- tectural features, and six photographs illustrating typical house styles. The RDG also describe eight commonly found house styles in Alhambra, along with their histories and typical construction materials, and are illustrated with photographs and sketches of design features. For example, Victorian style is illustrated by pictures of typical details such as pediments, turned balustrades, scalloped shingles, and bay windows with tall double-hung windows (Single Family Residential Design Guidelines 2009: 3–18, 3–19). While the RDG descriptions are meant to be informative to owners, designers and contractors, the language and photographs tend to be technical in nature and difficult to apply without expert or professional knowledge.

craftsman style

preferred building materials and colors

Wood Shake / Shingles *Wood Clapboard* *Stucco and Wood*

Stone (Accent Material) *Brick (Accent Material)* *Wood Trim (Painted as Accent)*

Most Craftsman homes utilize a primary exterior material (wood shakes, wood shingles, wood clapboard siding, and stucco) with an accent material such as stone or brick around the foundation, for supports, and for chimneys. These homes are often painted with a principal color and two complementary trim colors. Existing wood materials should be preserved and maintained. Stucco should be removed if not one of the original materials and should not be placed over existing wood siding.

preferred window and entry types

Picture Windows *Grouped Windows* *Paired Windows*

Windows on Craftsman homes are typically fixed, double-hung, and casement. A distinct Craftsman-style window has a diamond pattern or three, six, eight, or nine small panes over one large pane.

Glazed Doors (with Windows) *Paneled Doors* *Doors with Sidelights*

Craftsman doors are often wood with a stained finish. Windows within the door are arranged in distinct vertical and horizontal patterns. Battened, flush, and paneled types of doors are all appropriate.

Figure 19 Alhambra Residential Design Guidelines: Craftsman Style. Courtesy of the City of Alhambra.

By naming Alhambra's neighborhoods and identifying their borders, the RDG create and grant legitimacy to their history. The RDG also reference standard historic preservation resources in codifying house style attributes necessary to ensure adherence to stylistic "integrity." While city officials urge owners and designers to follow the guidelines in rehabilitation projects, they also use the guidelines to justify and authorize decisions made by the DRB or other staff. In fact, one of the primary reasons the city commissioned the guidelines in the first place was to provide support for dispute resolution and protect officials and board members in contentious decisions. Of course, the city recommends that applicants consult with the city architect or planning department staff before applying for any formal design review, but in practice few do. After the adoption of the RDG in 2009, however, the guidelines were rarely mentioned in any DRB hearings.

Design Review procedures require applicants to submit their home construction proposals to the planning department in advance of consideration by the DRB. The city's staff architect reviews all proposals in advance of the DRB meeting and writes an analysis of design qualities with a recommendation for DRB approval. Standard recommendation categories include "approval," "approval with conditions," "continuance," or "appropriate action" (implying a request for specific DRB design suggestions). Applicants, whether they are homeowners or designer-contractors, are expected to appear at the DRB hearing. The meetings have a formal quality to them in which the staff architect first reads his analysis and recommendation, followed by DRB discussion of the design's qualities, during which questions to the applicant may be posed before making a recommendation. Audience members may also be called on to voice their opinions about design proposals. It is in this context that APG members have an opportunity to raise issues related to the historic qualities of the houses that applicants propose altering. Since the adoption of the RDG in 2009, first time applicant approval rates have been improving, however, some applicants make numerous appearances before the DRB before winning approval for their designs.

DRB meetings

Before Alhambra's city council adopted the Residential Design Guidelines, the DRB was often confronted with challenges in communicating design ideas to lay homeowners and relatively unsophisticated construction professionals. Some of these issues are familiar in historic preservation communities, such as homeowners proposing to use stucco to cover wood-sided bungalows or replace wood-framed windows with vinyl ones, both prohibited practices. Other problems with outsized additions, or demolitions followed by new construction of large-scale houses, are common. Also, issues related to house style are frequent challenges. Some proposals are rejected because they lack an awareness of style, or included large rectangular shapes devoid of ornament. Other proposals mix stylistic elements or render details in an inappropriate manner. Even after the RDG were adopted problems continued, but at least now there existed a reference text to help guide applicants and legitimize the recommendations provided by the DRB.

Proposals to apply stucco to cover wood-sided bungalows, especially those that are classified as "Craftsman" bungalows, are notorious in many cities with preservation ordinances. In fact, it is one of the key remodeling outcomes that places an otherwise suitable older property beyond consideration as an historic resource, but many homeowners favor it as a quick, inexpensive modernizing fix for a deteriorated bungalow exterior. APG accounts of early DRB meetings make regular reference to homeowners' desires to apply stucco to their houses during the remodeling process. In 2006, one applicant wanted to stucco his 1922 Craftsman cottage to match the stucco on the new apartment units he proposed to build in the rear. An APG member in attendance commented that the audience was stunned, feeling they were witnesses to "a violent crime" that needed to be stopped. APG homeowners were encouraged to surveil their neighborhoods for inappropriate stucco and window replacement activity and report it. Although stucco advocates argue it is an economical means to enhance maintenance, such as painting, many homeowners actually prefer the aesthetic qualities of stucco. They seem to like the smooth surfaces and the more modern look the stucco achieves. Following on the denied application to cover his Craftsman home with stucco, one homeowner declared he would be looking for a different style to use for remodeling. Since 2006, the DRB has become firmer in rejecting stucco and window replacement proposals with the expectation that the practice would eventually decline.

Concerns about proposed changes in house sizes and scale were a second theme APG members reported in accounts of DRB meetings. Most of these applications sought to transform a bungalow into a mansion by demolition and new construction, by adding on a second floor, or by expanding the house footprint. One proposal sought to demolish a nondescript 1935 house and replace it with a 4,600-square foot, two-story home. The large one-half acre lot would accommodate the house, but the "quasi-Mediterranean" style was "too generic" to satisfy the DRB. Most of the projects the APG opposes, however, tend to be situated on smaller lots and involve adding a second floor that overwhelms the rest of the house. One particular proposal sought to double the size of the house from 1,156 to 2,737 square feet, but drew DRB criticism because the addition seemed "stuck-on" to the old design and its plan suggested an intention to divide it into two separate units. DRB advice focuses on stylistic detailing, especially when the additions look like plain boxes rather than being integrated into the original design. To help applicants reduce the visual appearance of overwhelming size DRB members suggest "massing" techniques that step back the second floor from the first and try to obscure second-story masses on the rear side of the house. The issues related to mansions and other oversized remodeling proposals have not diminished, although APG members feel their input has had a positive influence.

Concerns with stucco and mansion-sized remodels both draw attention to a third issue addressed in many DRB reviews, the integrity of design or "style." As a technical matter, design integrity refers to the stylistic consistency employed by the designer of any home project. Stylistic consistency presupposes a priori knowledge about "styles" or building fashions as historic artifacts, each characterized by a particular set of co-associated design features (e.g. roofline, door, window, etc.). Applicants who have little or no background in western design standards are particularly challenged by these presuppositions, an awareness of which drove the adoption of the RDG for educational and reference purposes during design review. The adoption of the codified guidelines, however, has brought to light the gap between the knowledge held by design experts and preservation homeowners, and the lack of this knowledge among the general public. The design guidelines emphasize a specific kind of cultural knowledge about historic styles that is more likely found among the middle classes and native-born citizens than among working-class residents or immigrants. For immigrants to succeed in acquiring an understanding of the appropriate design principles constitutes one way of passing the test of "cultural competence" (Ong 1996).

As early as 2007, APG members began noting that some proposals had been "continued" multiple times, having encountered tremendous difficulty winning DRB approval because of repeated stylistic difficulties. During the discussion phase of review meetings, board members typically asked applicants to explain their design decisions or made suggestions to help the applicant conform to a particular style they appeared to have in mind. If the design proposal was particularly difficult to interpret, a board member would ask if the applicant was "trying to design a Craftsman home" or inquire what style of home the applicant was trying to approximate in the design. The DRB message to applicants was clear, that they needed to pick a particular style, research its characteristics, and commit to following through with it in their project. Most important, applicants should resist the temptation to mix details from one style with another.

Once a draft copy of the RDG was completed in 2008, members of the DRB began recommending it to applicants and mentioned it in meetings: the problem was that many applicants, including property owners and designers, appeared unfamiliar with the idea of stylistic integrity the guidelines promoted, nor did they understand the application of principles to their projects. Some applicants became well known for their repeated appearances at the DRB with design issues that never seemed to get resolved. These applicants were unable to implement the style principles presumably due a lack of design sophistication and experience. While white and Latino applicants were equally challenged by the DRB's design expectations, they seemed able to learn. The largest number of the problematic encounters, however, primarily concerned the far more numerous Chinese immigrants.

In calendar years 2009 and 2010, a total of eighty-eight residential projects came before the DRB. These proposals included building separate units at the rear, or adding a second story, or a minor remodel of a porch, patio, façade or front yard wall. Proposals to add square footage amounted to 53 percent of total proposals in 2009, including 67 percent proposed by Asians, and 29 percent proposed by Latinos; and 70 percent of total proposals in 2010, with Asians proposing 77 percent and Latinos 17 percent; whites and other groups constituted the remainder. These percentages reflect growing size of the new Chinese immigrant community whose relative affluence and great interest in real estate investment aims to build wealth by appealing to local co-ethnics, including family, and transnational homebuyers.

In reviewing project proposals, Alhambra's staff architect frequently notes problems with "style" in his reports:

The proposal to extend and remodel the front façade of this home needs greater design attention. The existing tudor style home shows architectural distinction in its roof slope and the existing detail. The generic extension in both plan and elevation should be revised to better match the existing *architectural style and detail.* The new elevation should be of similar (or better) value and character as the existing. Recommendation is for continuance. (Alhambra Staff Architect's Report, October 2010 [emphasis added])

Or,

From the photographs the existing house appears to be a previously altered craftsman bungalow. The proposed remodel is by no means a restoration, however some of the alterations are sensitive to the original design intent of the residence. The use of a dormer and relevant attic eave vents are positive effects. The window type used should be re-considered as the elements that are chosen for the work would be a significant enhancement to the *perception of architectural style.* Recommendation is for appropriate action. (Alhambra Staff Architect's Report, November 2010 [emphasis added])

Presumably, many of these design problems could be corrected with additional meetings between the applicant and the staff architect, or with the applicant hiring a professional designer familiar with the city's design requirements. Some problems, however, lie with the designers or contractors themselves, who seem to be uneducated or naïve practitioners without much awareness of architectural design. One couple proposed to demolish their 1922 bungalow and replace it with a two-story home with a three-car garage. The proposed design included some "Craftsman-lite" characteristics applied as decorative elements. The DRB worked with the family and their designer to improve the Craftsman features of the home, but encountered resistance in the form of lack of comprehension. The APG member commented:

They and their architect cannot understand what a "Craftsman-styled home" means and had a design that was a basic "half-ass" Craftsman as Laura called it. It sort of resembles the Ralph's market on Huntington and Garfield. There was much discussion about the meaning of Craftsman and (the staff architect) said it would be more strategic … if they came up with a real Craftsman design to replace the significant house they were destroying. They were willing to try to do this, but their architect did not seem to understand something as basic as a true Craftsman window. The design had French doors and a concrete roof. (APG meeting notes, March 11, 2008).

Some applicants who attempt a do-it-yourself approach to home remodeling have neither the financial resources nor the knowledge to make proper "quality" repairs on their houses. A Latino homeowner, wanted to add a second floor and change the roofline of his house, which would effectively destroy the key elements of its Spanish style. At his second appearance before the DRB, the questioning probed his motivation for the design in order to advise him about a viable direction to take.

Board Member #1: (Pointing to one of the drawings.) The window, arch, and doorway you have don't go with the style you have. They are not consistent with the style.

Board Member #2: Were you trying to change the pitch of the roof from flat one to fix a leak?

Applicant: Yes.

Board Member #2: Or, were you trying to change the style of the house?

Applicant: No.

Board Member #2: Since the house you already have is Spanish, keep it Spanish. There are ways to make a pitched Spanish roof. Last time you took out the Spanish character and you were trying to make it Tudor, but the changes you propose will create more leakage problems.

Board Member #1: Just in case you are confused, last time we addressed the inconsistency of style. I'm not sure you understood the issue because the changes don't make sense. I'd advise you to keep the original Spanish style. If you like your neighbor's style (Tudor), we won't hold it against you, but whatever you choose, you must stick with it all the way through.

Board Member #2: I would recommend you get some professional help with the style. When people put in a replacement door using any door from Home Depot, you may not notice the change in the detail, but it makes a difference to the style.

(DRB Meeting, December 8, 2009)

One designer–contractor whose business involved him in multiple renovation projects made repeated appearances before the DRB and became somewhat notorious among board and APG members for his seeming inability to understand what was required. An APG member reported:

This is an application for a 2-story job so totally lacking in any kind of design, that the DRB had been sent back pleading for some type of architectural style—ANYTHING other than what it is—a characterless box. The DRB asked Mr. Tom Yee, the "designer" several questions such as, "Can you tell us what kind of design this is? What ties the first floor to the second floor?"—but it appeared that Mr. Yee could not understand what was being asked. Finally, he said that he would submit a "Spanish design" … It was continued. (APG meeting notes, May 13, 2008).

This same designer–contractor also encountered difficulties on other projects, each of which required five or more hearings at the DRB. One was a proposal for a second-story addition to a small Spanish-style home, which was presented and continued six times because the applicant was struggling to execute a suitable design. The DRB appreciated the designer's efforts because his first proposal for the same property rendered the house in a "pseudo-Craftsman" style with too much stucco, columns too thin, and an inappropriate window design, but he still had difficulties completing the Spanish design.

One of the more troubling aspects appearing multiple times in the same design proposal was an unusual number and configuration of interior spaces. Noting the two dining rooms and excessive number of bedrooms, one DRB member was able to discern that the floor plan would seem to allow for the later subdivision of the house into two units, a condition prohibited by the R–1 single-family residential zone in which the property was located. In another project that proposed two houses on a lot showed upper floor "wet bars" that were really full kitchens. The same DRB member spotted the deception and challenged the designer-contractor about why a single-family house needed two full kitchens. Clearly, the issues involved in some of the proposals went beyond the capacities of the designer-contractor himself and concerned the owner's interests in future development possibilities. These were not the only cases where interior plans suggested a secondary but somewhat hidden or even "subversive" rental agenda.

Homeowners' attempts to circumvent the rules and procedures may have expensive consequences. In 2008, Mrs. Lo, bought a 1913 Craftsman bungalow at the edge of an "historic district" that had been defined in the RDG and began to remodel it into a residence with a rental unit for her daughter. She hired a contractor who assured her that he had gotten permits for replacing the original wood siding with Hardiboard siding and replacing the original wood windows with vinyl ones. As the project was nearing completion, the neighbors and APG members complained to the city, which issued a stop-work citation to stop the work. Mrs. Lo and her husband were compelled to appear before the DRB in December 2009. The couple claimed their contractor had disappeared after spending most of their money on the remodel, but it was also evident that the contractor had not obtained permits, and now the couple needed to legalize the project. The staff architect reported, on December 8, 2009, that:

> This craftsman styled residence was originally clad in wood shingles and although the documentation is limited, the value of the original design was significant. Another key aspect of this application is the location of the house within a neighborhood/district of craftsman homes. The current owner has replaced the shingles and the windows with Hardiboard siding and white vinyl (or aluminum) windows. The residence has been cited and needs Board approval to legalize the new condition. Recommendation is for appropriate action.

Appearing at the hearing were some of Mrs. Lo's neighbors, preservationists, and a planning commissioner who wanted to speak of concerns that some members of the Alhambra community were trying to side-step city oversight. She argued that if approvals were given for the work already done, then others would be encouraged to do the same once word got out. Preservationists at the December 8, 2009, meeting were outraged by the remodeling practices the couple had used and said that given the position of the house marking the entrance to the historic district, "the residence … is a cheap cartoon imitation of a Craftsman—damaged and devalued." Although many homeowners seek project approval after construction has begun, the nature of this project underscored suspicions many have that immigrants had been avoiding getting required permits. Even the staff architect complained that the DRB was subverted when the owner and contractor failed to notify the city and it learned of it when neighbors called the city to halt the work. In the DRB meeting, December 8, 2009, he said that homeowners have the primary responsibility to get the proper permits, and lacking knowledge is not a valid excuse:

> We are seeing more and more of these types of projects. I think it is a serious matter for the DRB, this trend with certain unethical builders and designers for whom there aren't many repercussions after the fact. But the fact that it was done illegally is a serious problem. I hope word gets out that people won't get away with this gutting of old houses. We all worked so hard to get the design guidelines put in place.

Mrs. Lo eventually won approval from the DRB after promising to replace all the Hardiboard siding with new wooden shingles and the vinyl windows with wooden ones. Although she cried throughout the meeting saying that she had run out of money, the DRB was not moved by her emotional display. She found a new contractor who understood what was required and the project was finished by March 2010. As it neared completion, I stopped by to talk to her about the process. She told me that she did not mean to cause trouble by not getting permits, but felt betrayed by her first contractor. Still, she really did not understand what "historic" meant as she pointed to the remodeled two-story stucco house across the street. She said she had asked the neighbors if it was just her house that was historic, but they said it was the entire neighborhood. And, she added, that in the Philippines there are no regulations like these on houses. "We get permits, but they let us do whatever we please. Had I known it was historic, I never would have bought the house. Nobody told me," she said. By December 2011, Mrs. Lo had placed the house on the market.

Figure 20 Bungalow undergoing remodeling. Credit: Denise Lawrence-Zúñiga.

Homeowners' desires to remodel existing houses, or demolish them to build large new houses are not unique to Alhambra, but appear in neighboring cities throughout the ethnoburb. In fact, according to the staff architect, Alhambra's DRB itself was formed during the 1980s as a response to what locals saw as rampant "mansionization" in nearby Arcadia. Building mansions is not such a problem when there is abundant land for expansive ideas of home, the problems are more likely to surface where proposed remodels or new construction contrasts sharply with the neighborhood context most often filled with one-story bungalows. An associated problem that often occurs in large-scale proposals is a lack of stylistic development. Designs are said to appear "too heavy and massive," like two concrete boxes with details stuck on as an afterthought. Some DRB members have compared proposals to stucco tract-home construction, boring and lacking in any particular identity. And if the proposed new construction is simply a remodel of the original, a similar problem often rests with the lack of integration with the style of the existing house. Witness an APG report on one house:

> This house was built in 1927. It appears to be a Tudor cottage. The owner is proposing massive second story addition, which will more than double the square footage (from 1,156 to 2,737). Most of the homes in the neighborhood are much smaller with 2 or 3 bedrooms. The new house would have 6 bedrooms, 6 baths. The DRB members are very reluctant to approve this project. They've sent it back to the drawing board 4 times now. They've criticized the size of the addition, and expressed the opinion that it just seems "stuck-on" to the existing home, with virtually no integration of old and new designs. The have also been concerned about how easily this proposed design could, at some point in the future, be broken up into two separate units. The project was continued without approval. (APG meeting notes, April 26, 2005)

Of course, not all owners build large scale houses with the intention of subdividing spaces for future rental purposes, but often they do propose expanding the footprint of the house to cover as much of the lot as is allowed by the zoning ordinance (called Floor Area Ratio, or FAR). Aside from the lack of any distinguishable style, APG members complain about extensive lot coverage and a loss of "green space" in proposals that cover the yard with concrete. Local residents are also very vocal about the generally large size and scale of new two story houses that, in their minds, destroys the sense of neighborhood. They oppose the radical changes to residential landscapes that take no account of the existing historic context, but insert large stucco boxes into streetscapes that have maintained a smaller scale and settled character for some time. These new constructions are disruptive in size and assert an incompatible new aesthetic.

Sometimes, even when the DRB approves the design proposal, the Planning Commission, which is the next stop on the permitting process, denies approval. In one case, the Planning Commission later rejected a proposed two-story remodel, which had won approval from the DRB, for the same reasons that had concerned the board. It sent the project back to the DRB for redesign. The staff architect observes:

This application was approved by the DRB on September 16, 2006. The Planning Commission continued the application requiring redesign and it is being presented again to the DRB. Some of the concerns expressed by the Commissioners on the previously approved design include: *the lack of compatibility of the architecture style with respect to the neighborhood, the lack of architectural detail and a consideration given to different setbacks between the first and second floor.* The design presented at this time is the work of a different designer. The new design shows that while different setbacks has [*sic*] been incorporated, none of the other concerns identified have been addressed. Consideration should be given to engaging the services of an architect capable of executing the required architectural style, detailing and neighborhood compatibility. Recommendation is for continuance. (Alhambra Staff Architect, January 13, 2009 [emphasis added])

One member of the APG attending the same DRB hearing reported:

Both the homeowner and the contractor were angry about the forced delay of their project. They could not understand how they had met all the requirements of the DRB in 2006, but were subsequently rejected by the Planning Commission for design reasons. The contractor in particular was angry to the point of combativeness—he walked out of the meeting in frustration. (APG meeting notes, January 13, 2009)

In another case, the owner and designer made a proposal to the Planning Commission, in 2009, without ever having received approval from the DRB, but once it was turned down there on appeal by a unanimous vote (9–0) it returned to the DRB. The design was for a two-story addition to an existing Spanish Revival style house built in 1927 and was originally proposed to the DRB in 2008 on four separate occasions without receiving approvals. The Planning Commission reported that the design proposal introduced "detrimental alterations [to] the neighborhood character and stability; excessive bulk and massing; and lack of compatible proportion and scale with the surrounding area" (as quoted in APG meeting notes, February 23, 2010), and it sent it back to the DRB for reconsideration. Here again, the RDG played a significant role in helping the designer reshape the proposed project, and in early 2010 the DRB approved the project so that it could be presented to the Planning Commission.

APG members often argue that it is precisely these kinds of cases that benefit most from the Residential Design Guidelines. If owners and designers only knew what was expected, they reason, designs would conform and easily win approval. Throughout the first half of 2009, APG members worried about delays in getting city council approvals for the proposed guideline that had been presented some fourteen months earlier and since vetted by staff, boards and commissions. Once the guidelines were approved later in 2009, however, they had to be marketed. APG members often took it upon themselves to promote them, especially to applicants who had a difficult time winning approvals at the DRB meetings.

In one case in early 2010, an applicant had proposed a two-story remodel of a Tudor Revival house, which, according to the Staff Architect, was well executed and should be approved. An APG member who attended, however, drew attention to the proposed loss of the original distinctive rolled-edge roof for a new lightweight concrete tile roof. After some discussion the DRB withheld final approval. After the meeting, the APG member approached the owner to ask if she knew about the Design Guidelines and urged her to consult with the staff architect to get his input. In the following month, the owner returned with the revised design showing retention of the original rolled-edge roof and won immediate approval and the appreciation of the APG.

New ways of seeing

These design review events and outcomes suggest some broadly exclusionary processes at work in Alhambra under the banner of historic preservation values. The city's adoption of RDG has helped elevate awareness of preferred house styles and construction practices, and reaffirmed the DRB's role in insuring compliance. Both the APG and city anticipated that compliance would become less of a challenge over time as word about design review got out to the community. In 2010, the first year after the RDG were adopted, the number of first-time applicants winning approvals increased from 44 percent to 53 percent. If Asian surname applicant records are examined, that increase in approvals was from 45 percent in 2009 to 64 percent during 2010, a rise of almost 20 points. On closer examination it appears that many more owners began to hire design firms familiar with "mansion" design and the types of historic styles Alhambra promoted rather than change their own aesthetic preferences or work with local contractors used previously. That Chinese homeowners hire design professionals, just as homeowners in other ethnic groups do, suggests that they have learned an expedient way to pass the test of cultural competence (Ong 1996). Rather, the reworking of "self-making subjectivities" involves appealing to expert knowledge, as opposed to internalizing any aesthetic reform; but it does require recognition that designers have the expertise. The city's adoption of the new design guidelines has stimulated the growth of Asian-owned design firms who specialize in designing more palatable remodels and mansions.

It is unclear if or how design review as a form of aesthetic governmentaity has significantly altered homeowners' aesthetic perceptions or preferences. Ghertner argues that the visual techniques used in New Delhi are meant to reconstruct the subject's sense of self and place around a dominant aesthetic vision of the city and to encourage subjects to see themselves as the problem (2010: 207). Alhambra's design review, however, lacks the coercive intent Ghertner sees in New Delhi, but also produces myriad responses, which may be due to vast differences in subjects' wealth in the two settings. One Alhambra homeowner, whose initial proposal for a Mediterranean-style mansion was turned down by the DRB, hired a new designer who proposed the same square footage in a Craftsman-style design, which was accepted. When questioned, the homeowner admitted that he did not know the differences between the two styles or what the designer had done differently. But he offered a reason why the second design won approval, and that was because the DRB "was in a better mood." Clearly, he attributed some capriciousness to the DRB decision rather than credit the rational use of codified architectural styles.

The appeal of historic architectural styles among recent Chinese immigrants may be undercut by deeper counter perceptions and values. Interviews with some Chinese residents revealed strong cultural sentiments that militate against buying and restoring older houses. Some immigrants say that older houses are too costly to maintain; that because they are old and deteriorated, it is better to rip out the older features or demolish the structure and replace it with new construction. This is particularly true in kitchens and bathrooms, but also in expanding the often-cramped bedroom spaces and adding additional square footage to accommodate a family room. Some have suggested that the older houses are worn out, used up and should be discarded or completely renewed. One woman expressed disgust about her old house and another said the old wood in her house smelled badly, like the house she and her family had first occupied when they arrived in California (and were much poorer). Another mentioned that her old house reminded her that others had lived there before, people she could not know and did not want to know. She claimed to feel much better after the completion of a custom remodel with all the amenities, which brought her and her family a sense of pride. The rejection of old, antiquated houses and their replacement with new construction that owners can use to show off their new wealth and success is an important marker of social status.

Aspirations for newer and larger houses, including mansions, are not exclusive to Chinese immigrants, but are found among other affluent immigrants and native-born Americans of all ethnicities in southern California. Although Chinese immigrants are able to secure designers and contractors through their ethnic networks to design and build, and help them get government approvals, they also feel the pressure to respond to local and transnational co-ethnic buyers' demands for a modern design. Home buying and remodeling are often seen as investment opportunities, and Chinese investors have in mind the needs and desires of potential overseas buyers. In China, the number and wealth of the merchant and entrepreneurial classes have recently grown, and many aspire to come to California to live and invest their money in real estate. Zhang observes that because these people are not members of the elite classes, they feel insecure about their social status, and find consumption a reassuring outlet for demonstrating standing (2010: 131). Buying the right house in the right neighborhood with good schools signals entry into middle-class status, while all manner of consumer goods are marketed to appeal to Chinese aspirations to acquire Euro-American modernity. China's impatience with the old, and a strong desire to catch up with the West, has driven rampant construction in new housing schemes. These modern schemes demolish old neighborhoods, reinforcing consumption values favoring new commodities. Chinese immigrants living in Alhambra are not immune to these overseas influences, which they bring with them as core values guiding investments in southern California real estate.

Alhambra's neighboring ethnoburb cities, including San Gabriel, South Pasadena, San Marino, and Arcadia, also have influence on homebuyers' and investors' preferences. The cities have all experienced some threats to older residential neighborhoods from Chinese immigrants' remodeling and construction practices. Some cities had adopted historic preservation legislation, or more recently design guidelines, to reduce the impact, but only after significant unchecked mansion development had occurred.

These mansionized landscapes now constitute the reference points, the ideal models, for Alhambra's Chinese investors who seek to emulate and justify similar structures on their properties. Prospective investors sometimes claim that soon the entire region will be covered with new "Asian–Mediterranean" mansions. Nevertheless, Alhambra's DRB and the APG view their challenge as one of educating citizens of the entire city, while pushing for the adoption of stronger, historic preservation legislation. Regardless of whether individual homeowners' aesthetic preferences are changed by the existing regulations, the house remodels permitted under the design review process are more likely to result in styles that appear more "historical" and compatible with existing neighborhoods than if homeowners were granted the latitude they desire. They expect the resulting "durable" architectural landscape will reaffirm the legitimacy of those historic styles, and provide a set of different reference points for an alternative vision for future homeowners.

Alhambra presents a far more contested residential landscape than those found in the other cities in this study. In part because the timing—the initial resistance to preservation and then, belatedly, the arrival of affluent immigrants seeking to transform neighborhoods into their cultural ideal—has made city dynamics much different than those cities where preservation has long been the rule. Preservationist and Chinese homeowners present two contrasting and competing ideals of residential design and neighborhood landscape. One ideal embraces the past and seeks to restore its qualities to the original conditions, as much as that is possible. The other boldly asserts that what was built before is used, spent and out of date, and needs to be retired or thrown out and replaced with new, modern and updated amenities that bring housing into the twenty-first century. Both, however, express middle-class aspirational ideals aimed at transforming housing in such a way that expresses aesthetic values and validates social status while potentially displacing existing populations. These contested aesthetics of place reveal more sharply than the other cities the issues of returning old houses and neighborhoods to their original "whiteness" or substituting a new and different transnational aesthetic.

Conclusion

The dilemmas and challenges associated with remaking Alhambra's older residential neighborhoods have implications for the larger ethnoburb region where Chinese immigrants are settling and for older suburban cities in general. While many cities have welcomed investment for improving housing stock and revitalizing older residential zones, there is also increasing local resistance from advocates of historic preservation and homeowners who favor maintaining traditional styles. Regardless of whether these contested landscapes are significant enough as historic resources, or are necessary to the construction of urban identity, they are being reshaped by the competing interests of both immigrant and preservationist forces. Municipalities struggle to satisfy their citizens with formal constraints on designs and public reviews, but the foundation of these struggles lies in political processes that determine representation on city councils and commissions, which can sway in favor of one interest group or the other. In the latest election in Alhambra, APG members told of one Chinese candidate bussing in elderly voters to secure an influential position on the city council. Chinese immigrants have quickly figured out how to change the rules of the game in several surrounding cities as well. With competing visions of the older suburban city, the transformation of the landscape is just getting underway.

Toward an Anthropology of the Protected Suburb

Examining Alhambra's recent struggles with historic preservation challenges from new Chinese immigrants draws dramatic attention to the role regulations mandating particular aesthetic norms play in all the cities in this study. The contested perceptions and expressions of proper house design clearly vary among the diversity of suburban constituencies, from those who subscribe to a version of preservationist cosmology to those with very different ideas. Opponents of historic preservation include many immigrants of affluent and more modest means, but also native-born homeowners of all ethnicities and income levels. The struggle for any group to realize its aesthetic ideals fundamentally relies on the values and sentiments of homeowners who advocate or demure when these conflicts arise in the public sphere. And, while some municipalities have embraced preserving their architecture, others have held long-term permissive positions encouraging maximum development. At every level, from homeowners' private remodeling decisions and neighborhood social relations, to civic engagement and advocacy, the home's material qualities have the power to symbolize different values and lifestyles, and play a critical role in constructing the domestic moral project. All homeowners are likely attached to some set of material values, and some of these attachments may be emotionally quite strong. Aside from municipal regulations governing "tidiness" and "order" that keep trash off front yards, and lawns mowed, no other set of aesthetic values are systematically legitimized in southern California suburbs as those associated with historic preservation.[1]

The organization and integration of material culture into people's everyday lives have broad powers to motivate both private remodeling practices and civic preservation concerns. From the moment of initial nostalgic attraction to the physical features of an older home, to renovating, modernizing or restoring its original qualities, evolving perceptions and values of materiality guide the homeowner. The reciprocal effects of each remodeling act to reinforce the relationship a homeowner develops with the house co-produces identity and lifestyle. Encounters with the lived history of the house, discovering the identities and works of the original builder and first owners, confer a kind of agency to the material qualities of the house that speak to and inform homeowner decisions. Material agency reciprocally binds occupants to their houses and enhances their experiential understandings and interpretations of history, giving meaning to everyday lives. The homeowner's creative investment of their own agency in the restoration process extends the biographical lineage of persons in place, and links the homeowner with others in the neighborhood that promote stewardship of the residential landscape. In this way, historic home preservation is not a solitary act but is socially produced within a community of collaborators and advocates.

Homeowners' narrative accounts reveal, however, that there is no single preservation experience, but rather a variety of intensities and commitments to achieving a "purist" interpretation or an "authentic" restoration. Many homeowners such as those in Pasadena and Monrovia, and a few in Alhambra, tend to be more committed to a strict interpretation of historical restoration, even to the point of recreating some of the actual historic conditions of domestic life. Proportionately more homeowners in Ontario and Riverside, by contrast, hold to a more casual historic aesthetic, or practical interpretation that bends house form to contemporary needs.

A cosmology that conceives of history primarily as an architectural representation of the past relies on original materials and forms to facilitate those recollections. Apart from homeowners' embrace of historic restoration practices, the cosmology makes independent claims to legitimacy by assuring the authenticity of house form through a regulatory structure aimed at restoring a building's original qualities. Legitimacy claims are further encoded in scholarship and legal statutes, in institutional arrangements and professional organizations, and are produced and reproduced in ideological and normative contexts apart from any singular act of preservation. They are prominent in state, national and international architectural conservation programs, but preservation cosmology also plays an essential role in organizing communities and local level activities. Preservation homeowners appropriate and promote the cosmology as a set of normative principles for the neighborhood and for the entire city through local organizations and campaigns to educate the general public.

The cosmology, to which homeowners, city officials, designers and preservation professionals subscribe, is also used to explain and justify local aesthetic preferences and construction techniques in returning homes to their original conditions. By the same token, when a municipality officially adopts protective ordinances and accepts the cosmology as legitimate, it privileges some aesthetic expressions and their advocates, while excluding or rejecting others. Although adopting preservation laws goes a long way to ensuring the conservation or recreation of the idealized historic landscape, it does not guarantee it. Rather, any proposed transformation of the physical character of the residential neighborhood can threaten the ideal, which requires continuous homeowner vigilance, while contestations over domestic aesthetic expressions often lead to less than satisfying compromises.

Home preservation activities and public advocacy tend to intensify homeowners' commitments to achieving cosmology's ideal. That intensity is often expressed during various design review, historic preservation committee or city council meetings in impassioned pleas for saving the city's architectural patrimony. Advocates' appeals are directed not only at city representatives, but also at the applicants and owners who propose schemes preservationists find unacceptable. Some of their passion to convert non-believers exaggerates and dramatizes their position, and their allegations run the risk of seeming frenzied and furious. Yet, municipal governments that promote the historic brand appreciate the homeowners' participation in independently reviewing projects and proffering advice. In some cities like Pasadena, Monrovia, and Alhambra preservation work appears as more of a collaboration on mutually agreed-upon goals. In Riverside and Ontario where homeowner advocacy is muted, the city may appreciate the lack of engagement preferring to relegate preservation decisions to professional consultants, city officials and regulatory commissions.

Once municipalities adopt historic preservation regulations and put them into action, the power of domestic materiality plays out almost inconspicuously as an exclusionary force privileging particular aesthetic preferences. Each of the five cities examined in this study has had a different experience regulating historic preservation. Some cities found their historic neighborhoods increasingly gentrified as prior residents were pushed out or left. Middle-class families had originally built the older bungalows in Pasadena, Monrovia, and Alhambra, but after the war, when those families moved out to new suburbs, the houses came to be occupied by lower income renters and owners. Many times those latter residents came to be associated with crime or gang activity and were later supplanted by the younger more affluent gentry. Riverside and Ontario also experienced deterioration from disinvestment in some but not all of their older neighborhoods. Most, however, were not destabilized after the war, but maintained upper-middle and middle-class populations. Riverside began establishing some historic neighborhoods during the 1970s, which may have prevented deterioration from an influx of lower income residents. In Ontario, where homes in older neighborhoods, "below the tracks," lost their historic qualities due to "inappropriate" remodeling, the addresses were dropped from the survey list of historic resources. In all these cities, however, the advocacy for and adoption of exclusionary historic preservation regulations have had a continuing influence on changing neighborhood places in the suburban city.

Transforming the suburbs

The differences in intensity and spirit of preservation activity tends to reflect the character of the community, and also the current dynamics and direction of social change in all five of the older suburban cities. Suburban change depends on the intersection of complex dynamics including a city's history, demographic and income distributions, class and ethnic divisions, the presence of immigrant enclaves, and the real estate market. In fact, historic preservation may be more a reactive response to change, a nostalgic longing for a seemingly unchanging past, than an independent cause of it. While responding to these larger historical and socioeconomic questions is beyond the scope of this study, evidence suggests some patterns. Has the city remained relatively stable, at least stable in the historic districts where families have held on to their homes? Or, when "white flight" occurred in the postwar civil rights' period, did deteriorating housing stock and lower property values invite demolition and rebuilding or sales and rentals to lower income people that, in turn, invited later gentrification? Or, as in the case of Alhambra, did encouraging developer activities initially provide incentives for immigrants, some of whom are or have become relatively affluent, that, eventually, made "saving" neighborhoods a contest over competing aesthetic preferences?

None of the cities described in this study can be said to have completely resisted the demographic and economic changes to their residential neighborhoods over the course of the twentieth century; even the more stable communities have seen shifts in ethnic and class composition. Nor are residents in these five cities alone in perceiving potential threats to the physical makeup of residential neighborhoods, as many California suburban communities have witnessed the transformative effects of new homeowner construction. Rather, this study has described five different histories of regulatory structures and cultural sentiments related to domestic material values that have produced distinct outcomes. The differences in these structures and values hint at the different roles played by elites who exercise power directly and indirectly through local economic endeavors and community governments. In both Riverside and Ontario, political elites exerted continuous control over the historic preservation process in order to advantage their investment and development interests in the downtown district while protecting their own homes and neighborhoods from outside threats. Their efforts may be seen as protecting a particular social order, rather than giving way to more activist preservation sentiments. By contrast, Pasadena and, increasingly, Monrovia have accommodated to varying degrees the demands of advocates in preserving older residential neighborhoods, showing flexibility to accommodate a wider range of needs. Monrovia in particular continues to evolve as Asian immigrants have recently begun to mansionize older houses that are not protected by land marking; this has in turn intensified residents' concerns in the political sphere. Even Alhambra has made accommodations in favor of preservationist demands by issuing residential design guidelines and authorizing its Design Review Board to oversee their implementation. Although each city confronts the sociopolitical and economic challenges of preservation within its own boundaries, larger changes in the region also influence local urban policies and actions.

A recurring issue has focused on whether North American suburbs are largely middle-class white enclaves (Nicolaides and Weise 2006). In California, migrant settlers from the Midwest and Atlantic coast who initially constructed residential neighborhoods and civic cores around citrus groves in the San Gabriel and Pomona valleys certainly were ethnically white. They copied and interpreted architectural styles consistent with places where they had originated and employed craftsmen, also migrants, who knew well Euro-American building traditions. White migrants built the original suburban houses for white families, while local Mexican, Chinese immigrant or indigenous populations were relegated to separate neighborhoods and cities. As real estate speculation grew, exclusionary sentiments emerged in California and nationally as a way to protect property values. Exclusionary practices such as restrictive covenants and property deeds limited home sales to "Caucasians only," which further concentrated home ownership in the hands of whites. "Negroes," "Mexicans," and "Chinese" homebuyers were perceived as a threat to home values because they were symbolically associated with lower-income minority enclaves. These exclusionary sentiments based on societal norms were embedded in institutional practices that segregated residential neighborhoods, and included realtors, banks and government agencies such as the federal government, and municipal planning departments. Although the U.S. Supreme Court ruled in 1948 that racially exclusive restrictive covenants could no longer be enforced, racial segregation practices in home sales, discriminatory policies and home loans continued for decades (Nicolaides and Weise 2006).

Racial categories such as "White" or "Latino" or "Black" are considered by many scholars to be culturally constructed (Harrison 1995; Hartigan 1997), but they are also spatially constructed. European Americans employed the "possessive investment in whiteness" (Lipsitz 1995) to construct a homogenized white identity in the suburbs that excluded ethnic minorities. In the postwar period government agencies invested heavily in financing suburban development through exclusionary FHA loans and policies, and by extending roads and highways to the suburbs. Inner cities were targeted for urban renewal projects, which further stimulated white flight, but also produced higher density housing where minority populations were concentrated. Spatially, inner cities produced identities of color while new postwar suburbs produced white identity.

Older suburbs with early twentieth-century bungalows often encountered a fate similar to inner city areas where building stock suffered from disinvestment and properties were converted to rentals or sold to lower income homeowners. As new owner-occupants or renters remodeled and "modernized" their houses, they lost their original "historical" characteristics. City governments, however, often lacked the resources and, one could say, the will to govern these neighborhoods, and some such as those in Pasadena's Cottage Heights and Monrovia became gang and drug infested, high-crime neighborhoods. It would seem in retrospect that gentrification by historic preservation homeowners actually "saved" these neighborhoods from their own demise, as many preservationists claim, and brought these places back to life.

Although the twentieth-century suburb was ground zero for the production of whiteness in American society, the current role of architectural preservation practices that aims at restoring the original and authentic beginnings of the neighborhood is less clear. Without explicitly claiming as much, historic preservation advocacy and protective municipal ordinances work to rehabilitate or restore the "whiteness" of the suburban neighborhood by regulating the symbolic form and meaning of housing. Rehabilitating historical Euro-American architectural styles re-establishes the symbolic qualities of the original construction, but also, by implication, the ethnic and class status of those who originally built and owned houses in the suburb. The houses instantiate the particular "white" and middle-class social and cultural values to which so many, including immigrants and lower income families, aspire and desire to make their own. As this study shows, architectural aspirations associated with historic preservation do not actually require homeowners to be white. But homeowners must be comfortable with the aesthetic values represented in the restored Euro-American house styles and be willing to employ them. Indeed, many immigrants mark their assimilation into white America by buying or building a house in the suburbs, and meet the cultural "test" of acceptance by ensuring its conformity to neighborhood aesthetic norms. The preservation and restoration of older neighborhoods, however, is largely made possible by the coordinated efforts of government authorities and mostly middle-class residents, the same kind of actors who established the suburbs as "white" in the first place.

The historically preserved neighborhood continues to operate as an exclusionary "white" space, or "place," through the efforts of local community organizations and municipal regulations that seek and secure residents' compliance. In striving to create the ideal "historic" neighborhood, advocates and local officials endeavor to realize an enduring place, a heritage, protected by laws. Cities frequently promote their historic landmarks to define their identity and highlight these protected places as tourist destination sites. Historic districts are resources exploited in promoting a city's image, or "brand," as was seen in Pasadena's nomination of Cottage Heights for the National Register of Historic Places. All these cities with historic preservation laws promote their accomplishments in preserving residential neighborhoods, as well as important civic buildings, on websites and in promotional materials. Some, like Riverside and Monrovia, also promote their preservation professionals, including craftsmen and researchers, and local historians. Civic leaders know that the historic landscapes found in protected residential neighborhoods not only preserve or increase property values, but also act to recruit to the city new residents who aspire to the historic lifestyle and who are likely to be good stewards for the older property they acquire. In this sense, the branding and recruitment, while not being specifically aimed at white people *per se*, indirectly argues for middle-class responsibility in protecting symbolically Euro-American property values and community life. Moreover, the restored houses are seen as more than commodities. Gutting a landmarked residential structure and selling its fixtures online, as was described in Ontario, is considered a deeply dishonorable and shameful act.

In many ways, the historic preservation of a district remakes the moral qualities of the neighborhood, holding residents more accountable to each other for their own actions and the actions of their neighbors. Forming a protected historic district obligates all the residents to improving or maintaining the material qualities of history in their homes. While the residents in older and more stable residential neighborhoods, such as those of Riverside and Ontario, tend to have long-lasting social relations with one another, the formation of a protected historic district tends to reinforce those relationships making them even stronger and more exclusive. For newly gentrifying neighborhoods, forming a district can energize preservation sentiments and actions, linking advocates together in the social construction of an historic utopia, an idealized or nostalgic neighborhood. The new gentry become active civic ambassadors as they try to convince and educate their neighbors about the values of preserving and restoring old houses. The new gentry also tend to be the loudest complainers at city hall about violators they observe in their area as they try to remake the landscape that recalls the past. While the dominant quality of more stable neighborhoods tends to focus on retaining the traditional qualities that have always been in the neighborhood—putting the brakes on any "modernizing" changes—those of the newly gentrifying neighborhoods focus on imagining, restoring and purifying an idealized past. Efforts of the latter are geared to presenting the nostalgic historic imaginary in the form of houses and neighborhood that appear to have never been any different.

Beyond preserving California's historic suburbs

Coupling historic preservation or conservation practices with gentrification, the displacement of lower-income residents by the middle-class, is a worldwide phenomenon (Atkinson and Bridge 2005). Gentrification has occurred to varying degrees in all the cities described in this study, although not necessarily in the neighborhoods described. In similar ways, neighborhood gentrification began in major American cities in the 1970s, especially where "disinvestment" and low real estate values provided "opportunities" that attracted middle-class interests and could be paired with urban revitalization schemes. Early public reinvestment in housing primarily targeted older inner cities in the northeastern U.S., Western Europe, and Australia, and was followed in the late 1970s and 1980s by programs stimulating private investment and more recently in the 1990s with intensified state involvement (Hackworth and Smith 2001: 466–8). Gentrification, however, is not synonymous with historic preservation, although gentrification frequently follows a successful effort aimed at preserving historic architecture in areas of decline. As was shown in Alhambra and neighboring San Gabriel Valley cities, gentrification can also result from new homeowners' efforts to completely remake an aging suburban neighborhood by razing old houses to the ground and replacing them with new large mansions. The North American west coast from Vancouver to California has seen Asian immigrants increasingly settle in inner cities and suburbs, the latter most often associated with Asian–Mediterranean-style "mansionization." In both cases of preserving or replacing the old bungalows, however, lower income residents are literally pushed out. While this tends to affect renters disproportionately compared to homeowners, in California where low property tax rates for long-term homeowners are legally protected, even when real estate values rise, existing homeowners may benefit from that increase.

As in other parts of the country, the preservation of older residential architecture in California is often lent support by local and state government policies and programs, through zoning restrictions, historic preservation regulations or design guidelines, and property tax incentives (like California's Mills Act). Most of these services and incentives targeted for residential properties, however, are not included as part of urban revitalization programs, although similar incentives may be promoted in developing adjacent commercial zones. Rather, they are provided to homeowner groups who have themselves taken the "risk" of moving in and renovating an older house, and who collectively form a new constituency to demand additional services such as improved streets and sidewalks, trash collection services, and better schools. In the initial phases historic preservation protections may be established long after the gentry have already "turned around" the neighborhood; in Pasadena it took over ten years to get historic districts formally established through the preservation ordinance.

Many older urban and suburban communities throughout the United States have attracted new homeowners interested in historic preservation and neighborhood revitalization that displaces prior occupants. In Atlanta, Georgia, young professionals bought up older single-family homes in inner-city neighborhoods in the 1970s to preserve them and gentrify the neighborhoods. Similar efforts to "reclaim" the city by explicitly advocating historic preservation and designating historic districts have included the original neighborhoods and, more recently, have been extended to the streetcar suburbs beyond (Swope 2011). Portland, Oregon, has also seen gentrification coupled with extensive historic preservation protections occur in its charming older bungalow neighborhoods surrounding the city's center; these once housed significant numbers of African-Americans who have been pushed out to neighborhoods on the east (Theriault 2014). In both cases the gentrifiers demanded the city provide legal protections for their property investments as well as additional municipal and commercial services for the lifestyle to which the new residents aspired.

Some of the same processes that couple historic preservation with population displacement have been observed in dilapidated inner city zones in Europe. In fact, the term gentrification was proposed by Ruth Glass (1964) who observed middle-class professionals moving into working-class London in the post war period. They renovated Victorian houses and cottages using their own "sweat equity," which eventually displaced the original occupants. While the bulk of these early rehabilitations were done without direct government support, state and municipal governments soon saw the new homeowners and their efforts as vital to the redevelopment of downtown commercial zones and to improving the economic health of the city. These processes of gentrification have also been observed in other London neighborhoods (Butler and Robson 2003) and similarly large European cities such as Paris (Clerval 2008) or Rome (Herzfeld 2009); in the post-socialist cities of Berlin (Berndt and Holm 2005) and Krakow (Krase 2005); and beyond in Istanbul (Islam 2005), Tokyo (Fujitsuka 2005), and Melbourne (Shaw 2005). Theories of gentrification have characterized the motivations of gentrifiers in two ways: the production-side, which focuses on economic opportunity in investing in a deteriorated urban core (Smith 1996), or the consumption-side, which emphasizes the "back-to-the-city" appeal of inner-city living for the "new middle classes" (Ley 1996). Regardless of approach, displacement of the original population has been the typical result of rising property values.

More recently neoliberal policies have purposefully embraced transforming the built environment as a key to attracting people back to the city to live, but also to stimulate the growth of tourism and associated businesses. While some of the earliest efforts to reclaim a city's older neighborhoods may have certainly included the gentrification of charming older architecture, in reality reclaiming ancient buildings for contemporary reuse is a time-honored tradition throughout Europe. State involvement in conserving historic architecture as part of its patrimony has a deep history, even when efforts invested in renovating public buildings are not always compatible with gentrification or with accommodating residents or nearby businesses. In short, heritage conservation in Europe, and increasingly in developing countries, is often initiated by the state for the state, to promote urban and national identity, to attract tourists who will spend money and stimulate the growth of business, and to fill city coffers.

In the port city of Rethemnos, Crete, Herzfeld (1991) describes how homeowners who were not displaced attempted to cope with construction requirements in their historic zone. Most of the stone houses had been built over centuries by successive occupations of Venetians and Ottoman Turks, and exhibited distinct architectural details that the state envisioned to be worth conserving. Homeowners sought to remodel and expand their houses, and to incorporate modern amenities for their growing families. However, acquiring the proper permits from a variety of notoriously difficult state bureaucracies, including national archaeological and historic conservation agencies and municipal building and planning offices, that all seemed to work at cross purposes, drove homeowners to engage in nocturnal construction without permits to avoid attracting attention from the authorities (Herzfeld 1991). While homeowners were not necessarily displaced by the historic conservation regulations, they were tested by elaborate bureaucratic requirements.

Indeed, scholarship on the conservation of historic architecture in developed and developing countries tends to focus on the effects of state policies and regulations on local populations who live in and around the sites (e.g. Silva 2009, 2014). In the 1980s, a debate raged in many rural Portuguese agricultural towns over whether residents whose houses were located adjacent to patrimonial monuments would be permitted to replace aging wooden doors and windows with modern aluminum ones. Planners, town officials and historic preservation experts from Lisbon and Évora held public meetings in southern towns to explain how the aesthetic and practical qualities of the new silvery replacements that homeowners preferred conflicted with the aesthetic ideal the officials were trying to achieve. After much discussion, and as a form of compromise, permission was given to the homeowners to install aluminum replacements that were colored brown, green or other hues that wooden doors and windows would have traditionally been painted. But one of the effects of the public airing of these policy issues was the ready adoption by many of the same town residents of new wooden door and window replacements, the preferred aesthetic choice of conservation experts (Lawrence 1996). Those residents expressed a sense of pride when describing their choice of wooden replacements, eagerly exhibiting their new knowledge about heritage conservation and their contributions to it. By the late 1990s detailed rehabilitation studies were underway resulting in recommendations for aesthetic recuperation of the town's material attributes and economic revitalization with an emphasis on tourism.

These and other studies of historic preservation and its impact on local populations often focus on the negative effects of state imposed regulations, which tend to cause displacement. But every case of displacement resulting from state revitalization programs is not necessarily a case of gentrification, even as new residents move in to take the place of the previous ones. In recent decades India, China, and other Southeast Asian countries have pursued neoliberal policies emphasizing urban renewal as a vehicle to improve the local economy and urban image. In the process they have employed strategies of demolition and the construction of new and higher density multifamily housing. Many of these schemes depend on urban planning strategies employing the "master plan" to show the idealized outcome. In China, the logic of these projects often revolves around a cultural logic of "lateness" and sense of embarrassment about being perceived as backwards that impels the government to engage in large-scale modernization at great speed (Zhang 2006). These schemes tend to ignore not only the preservation of any historic architectural resources in the area, but also the residents and merchants who are displaced in the demolition of existing structures.

Some scholars use gentrification to classify these cases of urban development and subsequent displacement of people and argue that the theory should be amended to accommodate emerging examples (He 2012). In China, for example, residents of urban districts threatened with displacement have become increasingly aware that their property rights were recently amended in the constitution. They now assert themselves collectively in civic activism and protests, and file complaints and lawsuits against developers (Zhang 2006). Despite citizen action, however, development and displacement eventually takes place. On the other hand, arguments for conserving existing historic resources, such as Beijing's courtyard houses (*hutongs*), or Shanghai's row house neighborhoods (*lilong*), point to the recent adoption of historic preservation laws. While historic preservation protections might be thought of positively, Arkaraprasertkul argues that their use in selectively conserving Shanghai lilong properties also belongs to the same branding strategy that recognizes that some history is necessary to making the city attractive (2012: 151). What urban planners may miss, however, is that it is not just the buildings that need conserving, but the way of life, and that to conserve the way of life requires that the dilapidated conditions be brought up to current standards.

Attention to aesthetic considerations in judging an area "blighted" and deserving of demolition is mentioned infrequently or in passing. Harms (2012) describes considerations of beauty and fresh air used by officials as justification for "spatial cleansing" in plans to remake Ho Chi Minh City in Vietnam. The plans inevitably involved the eviction of residents from the site to make way for high-rise luxury housing and commercial facilities. Ghertner (2010), mentioned earlier, also describes aesthetic governmentality strategies employed by Indian officials in processes aimed at relocating slum residents in New Delhi to make way for urban revitalization projects. These tactics work on the intimate level of the subject's sense of self and place, of belonging, based on an imaginary vision for the future. They succeed when the subject leaves, either by force or choice, because they do not fit into the state's portrayal of the proper order of urban society (Scott 1998). In both India and Vietnam, revitalization projects displaced existing populations, but it is not clear whether these should be considered examples of gentrification.

In a recent critique of the increasing extension of gentrification theory to analyze displacement due to urban revitalization in developing countries, Ghertner (2014) argues that cities in the global South do not exhibit the same kinds of property, planning and legal systems as postindustrial geographies where gentrification theory originated. In India, the state is often involved in the violent dispossession of property for their public–private partnerships to rehouse slum-dwellers, but demolition of the slum tends to occur first before any kind of resettlement. The state's use of force for capital accumulation, or "accumulation by dispossession," is aimed at producing private property from public resources. Displacement by state appropriation of land to enable the construction of new projects for the middle classes, then, does not seem to fit with previous gentrification studies. In this way gentrification seems inappropriate for analyzing contexts that have not experienced prior disinvestment as in the West, have non-owner tenure arrangements, do not feature developments for higher and better use, and depend on noneconomic force to succeed (Ghertner 2014: 1568).

Renters and squatters who do not possess title to the land, are the segments of populations most harmed in revitalization efforts whether by real estate economics activated by historic preservation or by eviction (Herzfeld 2006). Homeowners seem to fare better, especially when the state desires cooperation and collaboration in revitalization efforts. In pursuing neoliberal incentives to improve the local economy and promote tourism, however, the heavy hand of local or national government agencies may impose aesthetic requirements on homeowners, including many who do not necessarily sympathize with the state's preservation goals. Aesthetic guidelines employed in the southern California suburbs teach or coerce homeowners to make their house renovations conform to the city's design ideal. These may show practical advice about window replacements, or recommend or require façade treatments in a particular district, but they are enforced through the issuance of permits or penalties. These regulations will not necessarily make historic preservation advocates of these homeowners, but it will not necessarily displace them either.

Stronger than regulations, however, is the hold that the historic imaginary has for homeowners who embrace the ideals and goals of the preservation cosmology and advocate for it in the civic arena. These efforts are largely associated with middle-class homeowners who rehabilitate old houses as an expression of their aesthetic and cultural preferences and lifestyles. And, while their efforts may sometimes displace lower income populations, especially the renters, their appropriation of a neighborhood does not necessarily result in gentrification as classically defined. Rather, local elite households may embrace historic preservation restrictions for their houses as a way to protect the neighborhood from incursions of new residents who hold to different aesthetic ideals. Or, as in the case of Alhambra, historic preservation values may be asserted to counter an equally powerful gentrifying force, the acquisition of properties by recent immigrants whose preference leans toward a modern or alternative aesthetic. According to preservation advocates, it is perhaps only in post colonial North America, where the depth of "history" is comparatively shallow, that developers, landlords and homeowners need to be reminded that the destruction or modernization of older housing stock destroys what little there is of an historic memory. Without historic preservation legislation or residential design guidelines, advocates believe that something is lost each time a new owner renovates an older home.

And yet, as advocates argue for their nostalgic vision of history, it is important to remember that historic preservation regulations, like all aesthetic policies and practices, operate as exclusionary principles. Gentrification may be one outcome of these policies, but it is not necessarily the only outcome. Gentrification may equally result from new homeowners remodeling homes to a different aesthetic standard. The core idea in recapturing and preserving history as romantically as it can be imagined, strives to reconstitute "whiteness" symbolically in the preserved and protected suburb. Preservation homeowners do not intend to discriminate against any ethnicity or class, they welcome them, but they do discriminate against aesthetic preferences that conflict with their own.

Historic Preservation Interview Questions

Could you describe your house for me?

1. What is the date of original construction?
2. Who was the architect or builder?
3. Who was the original client or owner of the house? How long did they live here?
4. What would you call the style of your house?
5. Do you know anything about the house's previous owners? How many were there and who were they? Have you been able to piece together a history of the house?

Could you tell me a little bit about your family's experience with this home?

1. When did you purchase the house? Was it the first home you bought? Where had you lived before?
2. Did you grow up in an older home?
3. What attracted you to this home? How important was location, price, size, appearance, neighborhood, city and/or its age?
4. Do you consider your older house "historic"? What is it that is historical about the house?
5. Who are the household members that reside in the home and how are they related? What kind of work do you do for a living? What is your level of education?

What kinds of improvements have you made to the home since you acquired it?

1. Could you describe the condition of the house when you first acquired it? Was it in "move-in" condition or did it need a lot of work?
2. What are some of the projects you have completed, or are in the process of completing? What motivated the projects? Who designed the changes and who did the work?

3. How did you learn about how to improve your home? Where do you get ideas and information for the improvement projects? Do you do any research when beginning a project? Do you exchange ideas or referrals to contractors with others? What kind and with whom?

4. Where are you most likely to search for and buy home improvement materials and supplies? Why-cost, convenience, appearance, durability, easy to install, etc.?

5. What is your view of modern conveniences and necessities in an older home?

6. How would you describe your main strategy or philosophy in improving your home—preservation, conservation or restoration, beautification or modernization, renovation, reproduction or sympathetic rehabilitation? Do you want to minimize or maximize the changes?

7. What features of your house do you find aesthetically pleasing or beautiful? How have you tried to complement or improve those features? What do you find beautiful about your home and why?

8. How would you characterize your relationship with your home? Investment, shelter, sacred place, or refuge? Do you consider yourself a steward or caretaker?

Now I'd like to ask you about your involvement with neighbors or other residents in your community in organizations and activities associated with historic preservation.

1. Do you know your neighbors? Have most of your neighbors lived in their homes for some time or is there a lot of turnover of houses on the street? Do you socialize with your neighbors at all?

2. Do you think your neighbors like and appreciate the work you have done on your house? Have they complemented you on the house?

3. Do you belong to or participate in a neighborhood organization dedicated to historic preservation? Do you attend meetings regularly? Do you participate in other events such as home tours?

4. What is your primary interest in participating—social, informational, political, professional, or other? Does your spouse or significant other also participate?

5. In your opinion, what are some of the most important goals of the neighborhood organization? Why did the organization form? What is the history of the organization as you know it?

6. What is your opinion of historic preservation efforts in your neighborhood? Are all or most of the neighbors in favor of preservation? Is there a sense of community in your neighborhood? Are there any conflicts over historic preservation? How has your neighborhood changed since you first moved in?

7. What are some of the biggest threats in your neighborhood to restoring the historic character? Consider developers, rentals, mansionization, stucco-ing over wood sided bungalows, tasteless modernizations, etc. What are the biggest obstacles that must be overcome? What is the best way to address these issues?

8. Do you participate in other community-wide historic preservation organizations in your city or other cities? What do you think are the benefits of these organizations for you? What are their significant contributions to preservation efforts?

9. What is your opinion of your city's activities in promoting preservation efforts and protecting historic buildings and districts? Are officials doing enough or too much? Are they easy to work with? What are the most significant conflicts and how do they deal with them?

Appendix B

Historic Preservation Questions for City Officials

Background:

What is your current position? How long have you been employed in this position by this agency? What does the position require you do?

What types of related positions have you previously held? Where and when (please describe how any of these involved work with historic preservation related tasks)?

What educational training do you have for this job? What experience prior to taking this job did you have?

How would you define historic preservation?

Describe HP activity in the city: What departments in the city have responsibilities related to historic preservation? What are the responsibilities? What types of actions are they expected to take and with reference to what kinds of issues?

How many staff have expertise or specialization in preservation. In what departments are they located?

What types of legislation does the city have to protect historic resources? What protections are there for residential properties?

Please comment on:

What are the differences between historic preservation in residential and commercial areas?;

Landmark building protections (how many designated local, state, federal historic landmarks?); are these concentrated in any particular areas?

Landmark district protections (how many designated districts); what are the differences between the districts in terms of wealth/class of original vs current owners, and in terms of current demands on city staff? Were all the districts self-organized by residents? What prompts residents to seek to establish them?

Historic resource surveys—how many? How extensive are these surveys? Who initiated them and when and why were the surveys done? Have any properties been lost or de-listed from original surveys?

Design guidelines (Secretary's Standards)

Mills Act (how many applicants/recipients—who and where?; knowledge about program—who finds out and how?);

Utilization of CEQA for historic preservation protection when landmark status doesn't exist?

What are the most successful protections? What are the most favored by residential property owners? What are the most contentious and why?

What types of actions do you and fellow staff get involved in? (Building permits and certificates of appropriateness, stop work orders, demo permits, nominations, etc.) What are most common neighborhood issues (across neighborhoods or in some?)?

What are the types of review procedures and what entities are engaged (commissions, committees and subcommittees).

What are some of the recent/notable cases involving residential properties and/or neighborhoods that present difficulties to city staff? Can you explain who is involved and what the challenges are?

In your opinion, what are the biggest challenges/threats to historic preservation in the city?

What is your opinion about the challenges historic preservation that each of the following pose to city planners? (developers, community/neighborhood historic preservation advocates, historic preservation advocacy groups, politicians—city council, and other city staff).

What are the city's challenges in educating a diverse citizenry about historic preservation?

What is your opinion about private property concerns expressed by homeowners?

In your opinion, what are the political tensions, controversies and conflicts that you have observed in the city over historic preservation issues? What kinds of issues pertain to residential neighborhoods?

What has been the long-term impact of historic preservation activity been on the city? What do you think will happen in the future?

How do you think your city rates in terms of neighborhood level historic preservation compared to other cities (for example?)? Do you regularly communicate with officials in other cities? Which ones?

In what city do you live? Do you own an historic home?

Historic Preservation Questions for HP Professionals in Public or Non-profit Agencies, and Advocacy or Community Organization Leaders

Background:

What organization are you with and in what capacity? How long have you been involved? How did you become involved in this activity? What is your educational background?

Please describe the work you currently do. What is your position on promoting historic preservation in your sphere of work? What is its value? Why is it important to society? What are the main obstacles?

What is your experience with historic preservation in residential settings? What are the different challenges in preserving individual properties as opposed to districts? How is this different/same as historic preservation of public buildings?

In your experience, who is most likely to support historic preservation in residential settings? Why do you think they support it?

Who is likely to be opposed? Why are people opposed? (lack of education, money, taste, cultural knowledge, or property rights and developer interests)

What do you think of city and other government officials in their roles dealing with preservation issues? What do you think of local politicians and how they handle preservation issues? What about developers and other investors from inside or outside the community?

What are the biggest threats to residential historic preservation activity? What is the most important solution or strategy for overcoming these threats?

Do you have any favorite cases that illustrate the challenges of historic preservation action? Can you please provide some detail in explaining them and what you learned?

In what city do you live? Do you own an historic home?

Notes

Chapter 1: Framing Preservation

1   The distinction of fifty years of age as "old" is a relative concept. In the United States, architecture that is fifty years or older is old enough to be considered for historic preservation protections. In western European countries, a building fifty years old would likely be considered still "new."
2   Although some examples of minority suburbs exist, the expansion of the postwar suburb is largely associated with the growth of white, middle-class communities (Nicolaides and Wiese 2006: 321–2).
3   Redlining is a practice that lending agencies used to exclude certain residential areas on a map from providing services. These areas typically included both middle- and lower-income families and were dominated by ethnic minorities, so that the lack of lending also contributed to disinvestment and the deterioration of housing stock.
4   My husband and I own a 1926 bungalow-style house, which we remodeled in 1999, but I confess that we never felt the same passion for the "original" materiality of our house that I observed among the many preservation homeowners I came to know in conducting this research.

Chapter 2: Discovering Material Agency: Making the Preservation Homeowner

1   The distinction between "designer" and "architect" is an important one in the discussion of historic properties. The term architect is understood to refer to someone who is professionally trained and employed as a building designer, typically a licensed architect. Most of the people engaged in historic preservation work as builders or contractors are not professionally trained as architects, nor are homeowners who engage in construction in their homes. The distinction is maintained in this narrative by the generic use of "designer" if no licensed architect is involved.

Chapter 3: Restoration Strategies, Imagining the Past, and Reconstructing Historic Meaning

1   The 1908 Gamble House was designed by Charles and Henry Greene, well-known California architects of the Arts and Crafts style, for David and Mary Gamble of Procter & Gamble fame. The Greene Brothers designed many houses in Pasadena and throughout southern California. The Gamble House is owned by the City of Pasadena, and operates as a museum of their work.

Chapter 4: Historic Preservation as Cosmology: Municipal Regulations and City Dynamics

1   Although many historic preservation professionals maintain that CEQA legislation should not apply to residential properties, several cities in this study acknowledged that they do consider the impact, or potential impact, of CEQA obligations to protect all cultural and historical resources. The extent to which they utilize CEQA guidelines for interpreting the significance of an individual property varies from city to city, however.

2   The designation of "Oak Streets" is a pseudonym.

3   Cottage Heights is a pseudonym.

Chapter 6: The Gentry Move In: Education, Reform, and Advocacy

1   This narrative is taken from the neighborhood website, which I am not identifying out of concern for confidentiality agreements with subjects.

Chapter 7: Immigrant Challenges: Communicating Preservation Values across the Cultural Divide in Alhambra

1   This chapter draws significantly on material previously published: "Bungalows and Mansions: White Suburbs, Immigrant Aspirations, and Aesthetic Governmentality," *Anthropological Quarterly* 87 (3) (2014): 819–54.

Chapter 8: Toward an Anthropology of the Protected Suburb

1   Even Asian preferences for house features consistent with the strongly held Feng Shui cosmological principles that have become increasingly popular with designers, do not always find support in local municipalities. In a series of events in Arcadia, a Chinese homeowner petitioned the city to have his address, which included the numbers "444," changed to a different number because, he argued, in Chinese the word for the number four sounds like the word for "death." He maintained that the perception about the number discouraged buyers for his home. While previously the city had accommodated many Chinese residents to make the address change by charging them for administrative costs, negative sentiments had recently increased about the local Chinese community and the city council rescinded the policy.

References

Ahern, L. (2001), "Language and Agency," *Annual Review of Anthropology* 30: 109–37.

Alhambra, California (2009), *Single Family Residential Design Guidelines*, Downtown Solutions, a division of Civic Solutions. Approved April 27, 2009, Ordinance No. 02M9–4531.

Alhambra, California, Design Review Board Agendas, Staff Architect's Reports: 2009–11. http://www.cityofalhambra.org/imagesfile/agenda.pdf (accessed monthly 2009–11).

Alhambra Preservation Group, Alhambra Design Review Board Meeting Notes, 2005–10. http://alhambrapreservation.org/alhambras-design-review-meetings/ (last accessed 10 September 2010).

Altman, I. (1975), *Environment and Social Behavior: Personal Space, Privacy, Crowding and Territory*, Monterey: Brooks Cole.

Altman, I. and M. Chemers (1980), *Culture and Environment*, Cambridge: Cambridge University Press.

Anderson, B. (2006), *Imagined Communities: Reflections on the Origin and Spread of Nationalism*, rev. edn, New York: Verso.

Appadurai, A. (ed.) (1986), *The Social Life of Things: Commodities in Cultural Perspective*, Cambridge: Cambridge University Press.

Archer, J. (2005), *Architecture and Suburbia: From English Villa to American Dream House, 1690–2000*, Minneapolis: University of Minnesota Press.

Arkaraprasertkul, N. (2012), "Urbanization and Housing: Socio–Spatial Conflicts over Urban Space in Contemporary Shanghai," in G. Barken (ed.), Aspects of Urbanization in China: Shanghai, Hong Kong, Guangzhou, Amsterdam: The University of Amsterdam Press.

Atkinson, R. and G. Bridge, G. (eds.) (2005), *Gentrification in a Global Context: The New Urban Colonialism*, London: Routledge.

Bachelard, G. (1964 [1958]), *The Poetics of Space*, trans. M. Jolas, Boston: Beacon Press.

Benson, M. (2013), "Living the Real Dream in la France Profonde?: Lifestyle Migration, Social Distinction, and the Authenticities of Everyday Life," *Anthropological Quarterly* 86 (2): 501–25.

Berndt, M. and A. Holm (2005), "Exploring the Substance and Style of Gentrification: Berlin's 'Prenzlberg'", in R. Atkinson and G. Bridge (eds.), *Gentrification in a Global Context: The New Urban Colonialism*, London: Routledge.

Birdwell-Pheasant, D. and D. Lawrence-Zúñiga (1999), "Introduction: Houses and Families in Europe," in D. Birdwell-Pheasant and D. Lawrence-Zúñiga (eds.), *House Life: Space, Place and Family in Europe*, Oxford: Berg.

Bourdieu, P. (1977), *Outline of a Theory of Practice*, trans. R. Nice, Cambridge: Cambridge University Press, (1972).

Bourdieu, P. (1984), *Distinction: A Social Critique of the Judgement of Taste*, trans. R. Nice, Cambridge, MA: Harvard University Press, (1979).

Brown-Saracino, J. (2010), "Social Preservationists and the Quest for Authentic

Community," in Brown-Saracino, J. (ed.), *The Gentrification Debates*, New York: Routledge.

Brown-Saracino, J. (ed.) (2010), *The Gentrification Debates*, New York: Routledge.

Bruner, E. (1994), "Abraham Lincoln as Authentic Reproduction: A Critique of Postmodernism," *American Anthropologist* 96 (2): 397–415.

Bruner, E. (2005), *Culture on Tour: Ethnographies of Travel*, Chicago: University of Chicago Press.

Bryson, B. (2010) *At Home: A Short History of Private Life*, New York: Anchor Books.

Buchli, V. (2013), *The Anthropology of Architecture*, London: Bloomsbury.

Butler, T. and G. Robson (2003), *London Calling: The Middle Classes and the Re-making of Inner London*, Oxford: Berg.

California Historical Building Code (2013), California Code of Regulations, Title 24, Part 8, California Building Standards Commission.

California State Office of Historic Preservation, Department of Parks and Recreation, Technical Assistance Bulletin #8, *User's Guide to the California Historical Resource Codes & Historic Resources Inventory Directory*, 2004.

Chung, B. (2011), *Exceptional Visions: Chineseness, Citizenship, and the Architectures of Community in Silicon Valley*, Doctoral Dissertation submitted to The University of Michigan.

Clarke, A. (2001), "The Aesthetics of Social Aspiration," in D. Miller (ed.), *Home Possessions: Material Culture Behind Closed Doors*, Oxford: Berg.

Clay, P. (2010 [1979]), "The Mature Revitalized Neighborhood: Emerging Issues in Gentrification," in L. Lees, T. Slater and E. Wyly (eds.), *The Gentrification Reader*, London: Routledge.

Clerval, A. (2008), "Les Anciennes Cours Réhabilitées des Faubourgs: Une Forme de Gentrification à Paris", *Espaces et Sociétés* 1/2008 (n 132–3): 91–106. http://www.cairn.info/revue-espaces-et-societes-2008-1-page-91.htm (accessed March 15, 2015).

Cohen, B. (1985), *The Symbolic Construction of Community*, London: Ellis Horwood Ltd and Tavistock Publications Ltd.

Cooper, C. (1974), "Home as Symbol of the Self," in J. Lang, C. Burnette, W. Moleski, and D. Vachon (eds.), *Designing for Human Behavior: Architecture and the Behavioral Sciences*, Stroudsburg: Dowden.

Cooper-Marcus, C. (1995), *House as Mirror of Self*, Berkeley: Conari Press.

Dant, T. (2005), *Materiality and Society*, Maidenhead: Open University Press.

Douglas, M. (1966) *Purity and Danger: An Analysis of the Concepts of Pollution and Taboo*, London: Routledge and Kegan Paul.

Dumke, G. (1943), "Colony Promotion during the Southern California Land Boom," *The Huntington Library Quarterly* 6 (2): 238–49.

Duncan, J. and N. Duncan (2001), "Aestheticization of the Politics of Landscape Preservation," *Annals of the Association of American Geographers* 91 (2): 387–409.

Duncan, J. and N. Duncan (2004), *Landscapes of Privilege: The Politics of the Aesthetic in an American Suburb*, New York: Routledge.

Eliade, M. (1959), *The Sacred and the Profane: The Nature of Religion*, trans. W. Trask, New York: Harcourt, Brace and World, (1957).

Favro, D. (1989), "Was Man the Measure?," in R. Ellis and D. Cuff (eds.), *Architects People*, Oxford: Oxford University Press.

Florida, R. (2010), *The Great Reset*, New York: HarperCollins Publishers.

Fujitsuka, Y. (2005), "Gentrification and Neighbourhood Dynamics in Japan: The Case of Kyoto", in R. Atkinson and G. Bridge (eds.), *Gentrification in a Global Context: The New Urban Colonialism*, London: Routledge.

Gable, E. and R. Handler (1996), "After Authenticity at an American Heritage Site," *American Anthropologist* 98 (3): 568–78.

Gable, E., R. Handler, and A. Lawson (1992), "On the Uses of Relativism: Fact, Conjecture, and Black and White Histories at Colonial Williamsburg," *American Ethnologist* 19 (4): 791–805.

Gell, A. (1998), *Art and Agency: An Anthropological Theory*, Oxford: Clarendon Press.

Gentilcore, R. (1960), "Ontario, California and the Agricultural Boom of the 1880s," *Agricultural History* 34 (20) 77–87.

Ghertner, D. (2010), "Calculating Without Numbers: Aesthetic Governmentality in Delhi's Slums," *Economy and Society* 39 (2): 185–217.

Ghertner, D. (2014), "India's Urban Revolution: Geographies of Displacement Beyond Gentrification," *Environment and Planning A* 46: 1554–71.

Ghertner, D. (2015), "Why Gentrification Theory Fails in 'Much of the World,'" *City* 19 (4): 552–63.

Gibson, J. (1979), *The Ecological Approach to Visual Perception*, Boston: Houghton Mifflin.

Glass, R. (1964), "Introduction: Aspects of Change," in Centre for Urban Studies (ed.), *London: Aspects of Change*, London: MacKibbon and Kee.

Green, H. (1998), "The Social Construction of Historical Significance", in M. Tomlan (ed.), *Preservation of What, For Whom? A Critical Look at Historic Significance*. Ithaca, NY: The National Council for Preservation Education.

Hackworth, J. and N. Smith (2001) "The Changing State of Gentrification", *Tijdschrift voor Economische en Sociale Geografie* 92 (4): 464–77.

Hall, E. (1966), *The Hidden Dimension*, Garden City, NY: Anchor Books.

Handler, R. and E. Gable (1997), *The New History in an Old Museum: Creating the Past at Colonial Williamsburg*, Durham, NC: Duke University Press.

Harms, E. (2012), "Beauty as Control in the New Saigon: Eviction, New Urban Zones, and Atomized Dissent in a Southeast Asian City", *American Ethnologist* 39 (4): 735–50.

Harris, D. (2006), "Race, Class, and Privacy in the Ordinary Postwar House, 1945–1960," in Schein, R. (ed.), *Landscape and Race in the United States*, New York: Routledge.

Harrison, F. (1995), "The Persistent Power of 'Race' in the Cultural and Political Economy of Race," *Annual Review of Anthropology* 24: 47–74.

Hartigan, J. (1997), "Establishing the Fact of Whiteness," *American Anthropologist* 99 (3): 495–505.

He, S. (2012), "Two Waves of Gentrification and Emerging Rights Issues in Guangzhou, China", *Environment and Planning A* 44: 2817–33.

Herzfeld, M. (1991), *A Place in History: Social and Monumental Time in a Cretan Town*, Princeton: Princeton University Press.

Herzfeld, M. (1992), *The Social Production of Indifference: Exploring the Symbolic Roots of Western Bureaucracy*, Chicago: University of Chicago Press.

Herzfeld, M. (2006), "Spatial Cleansing Monumental Vacuity and the Idea of the West," *Journal of Material Culture* 11 (1/2): 127–49.

Herzfeld, M. (2009), *Evicted from Eternity: The Restructuring of Modern Rome*, Chicago: University of Chicago Press.

Hobsbawm, E. and T. Ranger (eds.) (1983), *The Invention of Tradition*, Cambridge: Cambridge University Press.

Hoey, B. (2014), *Opting for Elsewhere: Lifestyle Migration in the American Middle Class*, Nashville: Vanderbilt University Press.

Hoskins, J. (1998), *Biographical Objects: How Things Tell the Stories of People's Lives*, New York: Routledge.

Hoskins, J. (2003), "Who Owns a Life History? Scholars and Family Members in Dialogue," in R. Waterson (ed.), *Southeast Asian Lives: Personal Narratives and Historical Experience*, Athens, OH: Ohio University Press.

Hoskins, J. (2006), "Agency, Biography and Objects," in C. Tilley, W. Keane, S. Kuchler, M. Rowlands, and P. Spyer (eds.), *Handbook of Material Culture*, Los Angeles: Sage.

Ingold, T. (2000), *The Perception of the Environment: Essays in Livelihood, Dwelling and Skill*, London: Routledge.

Islam, T. (2005), "Outside the Core: Gentrification in Istanbul," in R. Atkinson and G. Bridge (eds.), *Gentrification in a Global Context: The New Urban Colonialism*, London: Routledge.

Jackson, M. (1995), *At Home in the World*, Durham, NC: Duke.

Jessop, B. (2002), "Liberalism, Neoliberalism, and Urban Governance: A State–Theoretical Perspective", *Antipode* 34 (3): 452–72.

Jones, S. (2010), "Negotiating Authentic Objects and Authentic Selves Beyond the Deconstruction of Authenticity," *Journal of Material Culture* 15 (2): 181–203.

Kaufman, E. (1987), "Architectural Representation in Victorian England," *The Journal of the Society of Architectural Historians* 46 (1): 30–8.

Kershner Jr., F. (1953), "George Chaffey and the Irrigation Frontier," *Agricultural History* 27 (4): 115–22.

Kopytoff, I. (1986), "The Cultural Biography of Things: Commoditization," in A. Appadurai, (ed.), *The Social Life of Things: Commodities in Cultural Perspective*, Cambridge: Cambridge University Press.

Korosec-Serfaty, P. (1984), "The Home from Attic to Cellar," *Journal of Environmental Psychology* 4 (4): 303–21.

Krase, J. (2005), "Poland and Polonia: Migration, and the Re-incorporation of Ethnic Aesthetic Practice in the Taste of Luxury," in R. Atkinson and G. Bridge (eds.), *Gentrification in a Global Context: The New Urban Colonialism*, London: Routledge.

Langer, S. (1967), *Philosophy in a New Key: A Study in the Symbolism of Reason, Rite, and Art*, Cambridge, MA: Harvard University Press.

Lawrence, D. (1996), "Tourism and the Emergence of Design Self-Consciousness in a Rural Portuguese Town," in D. Pellow (ed.), *Setting Boundaries: The Anthropology of Spatial and Social Organization*, Westport, CN: Bergin and Garvey.

Lawrence-Zúñiga, D. (2010), "Cosmologies of Bungalow Preservation: Identity, Lifestyle and Civic Virtue," *City and Society* 22 (2): 211–36.

Lawrence-Zúñiga, D. (2014), "Bungalows and Mansions: White Suburbs, Immigrant Aspirations, and Aesthetic Governmentality," *Anthropological Quarterly* 87 (3): 819–54.

Lawrence-Zúñiga, D. (2015), "Residential Design Guidelines, Aesthetic Governmentality and Contested Notions of Southern California Suburban Places," *Economic Anthropology* 2: 120–44.

Legg, S. (2005), "Foucault's Population Geographies: Classifications, Biopolitics and Governmental Spaces," *Population, Space and Place* 11 (3): 137–56.

Lévi-Strauss, C. (1963), *Structural Anthropology*, trans. C. Jacobson and B. Grundfest Schoepf, New York: Basic Books.

Li, W. (2009), *Ethnoburb: The New Ethnic Community in Urban America*, Honolulu: University of Hawai'i Press.

Lipsitz, G. (1995), "The Possessive Investment in Whiteness: Racialized Social Democracy and the 'White' Problem in American Studies," *American Quarterly* 47 (3): 369–87.

Lorenzen, J. (2012), "Going Green: The Process of Lifestyle Change," *Sociological Forum* 27 (1): 94–116.

Lowenthal, D. (1985), *The Past is a Foreign Country*. Cambridge: Cambridge University Press.

Lung-Amam, W. (2013), "That 'Monster House' is My Home: The Social and Cultural Policies of Design Reviews and Regulations," *Journal of Urban Design* 18 (2): 220–41.

MacCannell, D. (1976), *The Tourist: A New Theory of the Leisure Class*, Berkeley: University of California Press.

Macdonald, S. (1997), *Reimagining Culture: Histories, Identities and the Gaelic Renaissance*, Oxford: Berg.

Maskovsky, J. (2006), "Governing the 'New Hometowns': Race, Power, and Neighborhood Participation in the New Inner City," *Identities: Global Studies in Culture and Power* 13 (1): 73–99.

McCracken, G. (1988), *Culture and Consumption: New Approaches to the Symbolic Character of Consumer Goods and Activities*, Bloomington: University of Indiana Press.

McLean, R. (1973), "Altruistic Ideals Versus Leisure Class Values: An Irreconcilable Conflict in John Ruskin," *Journal of Aesthetics and Art Criticism* 31 (3): 347–56.

Miele, C. (1995), "'A Small Knot of Cultivated People' William Morris and Ideologies of Protection," *Art Journal* 54 (2): 73–9.

Miller, D. (1987), *Material Culture and Mass Consumption*, Oxford: Basil Blackwell.

Miller, D. (1995a), "Consumption as the Vanguard of History," in D. Miller (ed), *Acknowledging Consumption: A Review of New Studies*, London: Routledge.

Miller, D. (1995b), "Consumption Studies as the Transformation of Anthropology," in D. Miller (ed.), *Acknowledging Consumption: A Review of New Studies*, London: Routledge.

Miller, D. (2010), *Stuff*, Cambridge: Polity.

Mitchell, K. (1993). "Multiculturalism, or the United Colors of Capitalism?", *Antipode* 25 (4): 263–94.

Mitchell, K. (2004), *Crossing the Neoliberal Line: Pacific Rim Migration and the Metropolis*, Philadelphia: Temple University Press.

MOHPG (Monrovia Historic Preservation Group) (n.d.) "Our History", http://www.mohpg.org/mohpg-history.html (accessed August 13, 2015).

Monrovia, California (1994), Municipal Code, Zoning, Title 17, Chapter 17.40.060. http://www.amlegal.com/nxt/gateway.dll/California/monrovia/monroviacalifornia codeofordinances?f=templates$fn=default.htm$3.0$vid=amlegal:monrovia_ca (last accessed September 2, 2015).

Munn, N. (1986), *The Fame of Gawa*, Cambridge: Cambridge University Press.

Nicolaides, B. and A. Wiese (eds.) (2006), *The Suburb Reader*, New York: Routledge

Oliver, P. (1987), *Dwellings: The House Across the World*, Oxford: Phaidon.

Ong, A. (1996), "Cultural Citizenship as Subject-Making: Immigrants Negotiate Racial and Cultural Boundaries in the United States," *Current Anthropology* 37 (5): 737–51.

Pasadena, California, Historic Preservation Commission, Agenda Report, April 25, 2005.

Pasadena, California, Department of Planning and Community Development, Planning Report, April 18, 2005.

Press-Enterprise (Riverside), May 15, 1984; June 21, 1984; September 10, 1986; February 18, 1988.

Rapoport, A. (1969), *House Form and Culture*, Englewood Cliffs NJ: Prentice-Hall.

Redfern, P. (2003), "What Makes Gentrification 'Gentrification,'" Urban Studies 40 (12): 2351–66. http://usj.sagepub.com/content/40/12/2351 (accessed February 1, 2003).

Riverside, California (2003), Historic Preservation, Citywide Residential Design Guidelines, 2003. https://www.riversideca.gov/historic/guidelines.asp (accessed August 23, 2007).

Riverside, California (n.d.), Historic Preservation, Historic Districts and Buildings. Oak Streets Historic District. http://olmstedriversideca.gov/historic/dist_mtp.aspx?dky=15 (accessed August 27, 2007).

Riverside, California (2006 [1969]), Municipal Code, Cultural Resources Ordinance, Title 20, Chapters 20.05–20.45.

Rose, N. (1996), "Governing 'Advanced Liberal Democracies,'" in A. Barry, T. Osborne, and N. Rose (eds.), *Foucault and Political Reason: Liberalism, Neo-Liberalism and Rationalities of Government*, Chicago: University of Chicago/UCL Press.

Ruskin, J. (1989), *The Seven Lamps of Architecture*, New York: Dover Publications. Originally published 1880.

Rybczynski, W. (1986), *Home: A Short History of an Idea*, New York: Viking.

Shaw, K. (2005), "Local Limits to Gentrification: Implications for a New Urban Policy," in R. Atkinson and G. Bridge (eds.), *Gentrification in a Global Context: The New Urban Colonialism*, London: Routledge.

Shove, E. (2003), *Comfort, Cleanliness and Convenience: The Social Organization of Normality*, Oxford: Berg.

Silva, L. (2009), "Heritage Building in the 'Historic Villages of Portugal': Social Processes, Practices and Agents," *Journal of Ethnology and Folkloristics* 3 (2): 75–91.

Silva, L. (2014), "The Two Opposing Impacts of Heritage making on Local Communities: Residents' Perceptions: A Portuguese Case," *International Journal of Heritage Studies* 20 (6): 616–33.

Smart, A. and J. Smart (1996), "Monster Homes: Hong Kong Immigration to Canada, Urban Conflicts and Contested Representations," in J. Caulfield and L. Peake (eds.), *City Lives and City Forms: Critical Research and Canadian Urbanism*, Toronto: Toronto University Press.

Smart, A. and J. Smart (n.d.), "Urban Displacement in Hong Kong: Outcomes, Mechanisms, History and Emic Accounts." Manuscript.

Smith, N. (1996), *The New Urban Frontier: Gentrification and the Revanchist City*, London: Routledge.

Smith, N. (2002), "New Globalism, New Urbanism: Gentrification as Global Urban Strategy," *Antipode* 34 (3): 427–50.

Smith, N. (2010), "Toward a Theory of Gentrification: A Back to the City Movement by Capital, not People," in J. Brown-Saracino (ed.), *The Gentrification Debates*, New York: Routledge.

Spain, D. (1992), *Gendered Spaces*, Chapel Hill: University of North Carolina Press.

Stansky, P. (1985), *Redesigning the World: William Morris, the 1880s, and the Arts and Crafts*, Princeton: Princeton University Press.

Swope, D. (2011) *Quantifying Neighborhood change in Little Five Points*, Applied Research Paper, for Dr. Harley Etienne, Georgia Tech University. https://smartech.gatech.edu/bitstream/handle/1853/43477/AndrewSwope_Little%20Five%20Points%20Neighborhood%20Indicator%20Explanation.pdf (accessed March 15, 2015).

Tambiah, S. (1985), *Culture, Thought, and Social Action: An Anthropological Perspective*, Cambridge, MA: Harvard University Press.

Theodossopoulos, D. (2013), "Laying Claim to Authenticity: Five Anthropological Dilemmas," *Anthropological Quarterly* 86 (2): 337–60.

Theriault, D. (2014), "Here's What Four Decades of Gentrification in North and Northeast Portland Looks Like," Web blog post. History, Portland, City Hall, Build Out. Portland Mercury. September 25, 2014. http://www.portlandmercury.com/BlogtownPDX/archives/2014/09/25/heres-what-four-decades-of-gentrification-in-north-and-northeast-portland-looks-like (accessed February 23, 2015).

Tilley, C, (2006), "Objectification," in C. Tilley, W. Keane, S. Kuchler, M. Rowlands, and P. Spyer (eds.), *Handbook of Material Culture*, Los Angeles: Sage.

Thompson, M. (1979), *Rubbish Theory: The Creation and Destruction of Value*, Oxford: Oxford University Press.

Tobey, R. (1996), *The New Deal and the Electrical Modernization of the American Home*, Berkeley: University of California Press.

Tomlan, M., (ed.) (1998), *Preservation of What, For Whom? A Critical Look at Historic Significance*. Ithaca, NY: The National Council for Preservation Education.

Turner, V. (1969), *The Ritual Process: Structure and Anti-Structure*, Chicago: Aldine.

Tyler, N. (2000), *Historic Preservation: An Introduction to its History, Principles, and Practice*, New York: W. W. Norton and Co.

U.S. Department of the Interior, National Park Service, Preservation Assistance Division. *The Secretary of the Interior's Standards for Rehabilitation and Guidelines for Rehabilitating Historic Buildings*, rev. edn, Washington, DC: U.S. Department of the Interior, 1992.

Van der Hoorn, M. (2009), *Indispensable Eyesores: An Anthropology of Undesired Buildings*, New York: Berghahn Books.

Van Gennep, A. (1960), *The Rites of Passage*, trans. M. Vizedom and G. Caffee, Chicago: University of Chicago Press.

Veblen, T. (1989 [1899]), *The Theory of the Leisure Class*, Amherst, NY: Prometheus Books.

Weber, M. (1948), "Class, Status, Party," in H. Gerth and C. Mills (eds.), *From Max Weber: Essays in Sociology*, London: Kegan Paul.

Weiner, A. (1992), *Inalienable Possessions: The Paradox of Giving-While-Keeping*, Berkeley: University of California Press.

Williams, B. (1988), *Upscaling Downtown: Stalled Gentrification in Washington*, Ithaca, NY: Cornell University Press.

Wilson, D. (2004), "Making Historical Preservation in Chicago: Discourse and Spatiality in Neo-liberal Times," *Space and Polity* 8 (1): 43–59.

Wolf, E. (1999), *Envisioning Power; Ideologies of Dominance and Crisis*, Berkeley: University of California Press.

Wyly, E. and D. Hammel (2005), "Mapping Neo-liberal American Urbanism," in R. Atkinson and G. Bridge (eds.), *Gentrification in a Global Context: The New Urban Colonialism*, London: Routledge.

Zhang, L. (2006), "Contesting Spatial Modernity in Late-Socialist China," *Current Anthropology* 47 (3) 461–76.

Zhang, L. (2010), *In Search of Paradise: Middle-Class Living in a Chinese Metropolis*, Ithaca, NY: Cornell University Press.

Zukin, S. (1989), *Loft Living: Culture and Capital in Urban Change*, New Brunswick, NJ: Rutgers University Press.

Index

This index covers all chapters of the book. It categorizes features of Californian aesthetics, authenticity, historic preservation and homes, primarily in California; these principal topics and their subheadings are further categorized by other headings and subheadings. Abbreviations appear both in full and by abbreviated form. Terms are in full (abbreviations in parentheses). An "f." after a page number indicates a figure.

aesthetics 1, 15, 53–4, 58, 169
  aesthetic governmentality 141–3, 144, 145f., 146, 155
  agency 8
  anti-mechanization and 81–3
  disparity 4, 106
air conditioning 60
Alhambra 18–19, 28, 43, 54–5, 139–40, 155, 162
  cabinet 58–9
  challenges 101, 139, 140, 146, 157–8
  consumption 156
  disparity 140, 143–4, 155, 157
  door 32
  experimentation 53
  fireplace 61
  gentrification 161
  hidden agendas 151
  knowledge of styles and 147–50
  large-scale projects and 140–1, 147, 153–4, 156–7
  mistreatment and 143
  multiple occupancy 38
  orchards 18
  paint 38, 55
  retrospective permits and 151–2
  roofs 154–5
  scope 101–2, 140, 143, 144, 145f., 146, 148, 155, 156
  shingles 151
  stucco and 147
  tiles 43
  windows 43–4, 151
Alhambra Preservation Group (APG) 101–2, 139–40, 147

disparity 140
  knowledge of styles and 148, 149, 150, 154–5
  large-scale projects and 147, 153, 154
Altman, I. 6
aluminum 43, 167
Armsley Square 119
APG, see Alhambra Preservation Group
Arts and Crafts Movement 81–3
asbestos 40
Asian–Mediterranean property 143, 157
Atlanta (GA) 165–6
authenticity and originality 3, 11–12, 13–14, 33
  disparity 48
  essences and 12
  limitations 12

Batchelder tiles 61
bathrooms 43, 57–8, 60
Bedford (NY) 14

cabinets 28–9, 30f.
  costs and 58–9
  experimentation 72
  toe-kicks and 65f.
California 1–2, 19f.
California Environmental Quality Act (CEQA) 88
carbon monoxide leak 44
carpentry 54–5, 56, 59, 72
Certified Local Government (CLG) program 89
CEQA (California Environmental Quality Act) 88

Chemers, M. 6
childhood 7, 8, 23, *see also* family;
    memories
CHNA, *see* Cottage Heights
    Neighborhood Association
Claremont 71–2
class 105, 163–4, 169
    consumption and 9, 83
    districting and 90
    exclusion and 17, 113, 114, 161, 162
    heirlooms 9
    knowledge of styles and 148
    see also elites; gentrification
CLG (Certified Local Goverment)
    program 89
Colonial Revival property 71–2
consumption 9, 10, 81, 83, 156
coop 37
cosmology
    challenges 81, 104
    constrait and exclusion 108 (*see also*
        elites)
    "integrity" and 86, 87
    legislation and regulations 11, 14, 102,
        160
    power and 80, 102
    sacred propositions and 80
    scope 3, 11, 80, 81, 82, 102, 122, 160
    secular 11, 80
    "significance" and 85–7, 100–1
    *see also individual place names and
        local organizations*
costs 23, 27, 52–3, 55, 95
    quality work and 55, 58–9
Cottage Heights 135
    challenges 132, 133
    demolition and 133
    disparity 134, 137
    events and tours 130–1, 131f.
    gentrification and 132, 135–6
    large-scale projects and 132–3
    mistreatment and 131–2
    scope 130, 134–5, 136–7
    weekend work 133
Cottage Heights Neighborhood
    Association (CHNA) 130, 134–5
    challenges 132, 133
    disparity 132, 134, 136
    events and tours 130–1

large-scale projects and 132–3
    mistreatment and 131–2
Country Club 90
Craftsman property 3f., 28, 35, 42, 48
    air conditioning 60
    bathroom 43, 57–8
    carbon monoxide leak 44
    coop 37
    demolition and 126–8, 133
    disparity 27, 128
    doors 32, 37
    exclusion and 128
    experimentation 52, 71–3
    fireplaces 61, 62
    floors 52–3
    ghost marks 44
    heirlooms 47
    infestation and 127–8
    kitchens 62–3, 64, 66–7, 72
    knowledge of styles and 149
    memories 46–7, 75, 77
    mistreatment and 40, 64
    newspaper in wall 34
    original style and 34–5
    paint 38–9, 53
    picture rails 68
    purist approaches 56, 77
    retrospective permits and 151–2
    scope 145f.
    shingles 151
    stucco and 39–40, 117, 118, 147
    trust and 59
    wainscoting 35
    wallpaper 68
    weekend work 133
    windows 43, 151
crime 112–13, 136
Cultural Heritage Board (Riverside) 89,
    110
    challenges and 89, 91–2, 115
    limitations 88–9

deck 125
Design Review Board (DRB) (Alhambra)
    101, 144, 146, 147, 148
    challenges 146
    disparity 155
    hidden agendas and 151
    knowledge of styles 147–8, 149–50

large-scale projects and 147, 153–4
  retrospective permits and 151–2
doors 31f., 32, 37, 167
  purist approaches 56, 73–4
  stucco and 41f.
DRB, *see* Design Review Board
drugs 136
Duncan, J. 14, 106
Duncan, N. 14, 106
Dutch Revival property 27

eBay fixtures sale 119
education 47, 54–5, 82, 89–90, 107, 132
  events and tours 110–11, 123–4, 129,
    130–1, 131f.
  limitations 14–15
  parking issues and 111–12
elderly homeowners 24, 26, 54, 73, 78
elites 80, 108, 120, 162
  consumption and 10
  demolition and 129
  disparity 119, 134
  exclusion and 85–6, 90–1, 92, 106, 116,
    119–20
  mistreatment and 104
ethnicity 19f., 83, 140, 142, 143, 148, 155,
    156, 163–4
  challenges 140, 142, 157–8
  consumption 156
  de-listing and 94
  demolition and 126–8, 167
  disparity 17, 128, 135–6, 143–4, 156,
    157, 159, 161–2
  districting and 90
  exclusion and 16–17, 113, 114, 128,
    162–3, 167–8
  knowledge of styles and 32, 148,
    149–50
  large-scale projects and 140–1, 156–7
  mistreatment and 143
  retrospective permits and 151–2
ethnoburb 19f., 140, 143
  challenges 157–8
  large-scale projects and 156–7
  see also ethnicity
experimentation 4, 52, 53, 71–3
  carpentry 72

family 24, 28, 38, 40, 48

anecdotes 47
demolition and 129
disparity 24, 26, 27, 156
exclusion and 115
habitual use 6
heirlooms 47
hidden agendas 151
kitchens 74
memories 27, 46–7, 75, 77
purist approaches 73–5, 77
fir (Douglas) 52–3, 57–8
fireplaces 29f., 60–2, 68
floors 52–3
French Provincial property 73

Gable, E. 12
garages 39f.
Gell, A. 8
gender 6, 59, 90, 103
gentrification 1, 18, 163, 169
  demolition and 168
  disparity 107, 132, 135–6, 168
  drugs and 136
  events and tours 129
  exclusion and 15, 16–17, 18, 106,
    137–8, 161, 165, 166
  large-scale projects and 129
  profits 132
  scope 16, 121–2, 128–9, 135, 164,
    165–6
  see also ethnicity
Ghertner, D. 15, 141, 142, 155, 168
ghost marks 44, 45f.
Gibson, J. 6
Greene & Greene 98
grove property 110

habitus 7, 23, 24
Handler, R. 12
Hardiboard 151
heirlooms 9, 47
Herzfeld, M. 166
Historic Oak Streets Association (HOSA)
    111–13
Historic Ordinance Committee
    (Monrovia) 96, 124
historic preservation 1, 3–4, 9, 11, 52, 107,
    160–1
  agency 80–1, 82

challenges 81, 82, 139, 166–7
cosmology 3, 11, 80, 81, 82, 102, 104,
    122, 160
demolition and 84, 103
disparity 17–18, 23–4, 83, 122, 159,
    168
exclusion and 14, 18, 105–6, 169
"integrity" and 86, 87
legislation and regulations 1, 3–4,
    12–13, 15, 19, 84–5, 86, 159, 161,
    165, 167, 169
local concerns and 84–5, 86, 102, 103,
    104–5
mistreatment and 107–8
naturalized environments 8
scope 79–80, 83–4, 85, 105, 161, 164,
    165
"significance" and 85–7, 100–1
taste 105
urban renewal and 103–4
Holliston Avenue 99–100
homes 5, 10–11, 23, 25–6, 36, 152f.
    agency and 24–5, 48, 49–50, 51–2, 66,
        68, 70, 159–60
    celebrities 25, 33–4, 35, 44, 46
    challenges 78
    comfort 67
    disparity 159
    domestic materiality 1, 2, 4, 24
    first encounters 26–7, 28–9, 32–3
    habitable space 25
    lifestyles 10, 48, 49, 104
    mystery 26, 33
    objectification 7–8, 10
    original style and 33, 35–6, 42–3, 49, 50
    personalization 5, 8, 71, 79
    previous owners 33–4, 36, 37–8, 44,
        46, 66
HOSA (Historic Oak Streets Association)
    111–13

iceboxes 7–8
immigration, *see* ethnicity

jalousie windows 43
Johnson, John K. 35
Jones, S. 12, 13

Kawa Market 113

Kieft and Hetherington 34–5
kitchens 63–4, 66–7, 73
    cabinets 65f., 72
    experimentation 72
    mistreatment and 62–3
    original style and 63, 64–5
    porches and 65–6
    purist approaches and 60–1, 62, 67, 74
    refrigeration 7–8, 57f., 63
    stoves 63, 66
    wainscoting 72

libraries 88, 103, 111–12, 117
light switches 70

McCracken, G. 9
McLean, R. 82–3
mahogany 58–9
mansionization 126, 132–3, 140–1, 141f.,
    153–4
    challenges 154, 156–7
    plainness and 147, 153
markets 8–9, 15, 81
    anti-mechanization and 81, 82–3
    gentrification 16
    see also consumption
Mediterranean Revival property
    exclusion and 115
    experimentation 53
    fireplace 61
    trust 59
    windows 43–4
memories 6, 27–8, 46–8
    challenges 75
    romanticization 26–7, 77
Miele, C. 83
Miller, D. 7, 10
Mills Act 88
Mission Inn 109–10
Monrovia 18–19, 28, 32, 48, 54, 96, 123,
    124, 162
    challenges and 97–8
    deck 125
    demolition and 97, 126–8
    disparity 26, 128, 137
    districting and 96–7, 98, 124–5
    doors 37, 73–4
    events and tours 123–4
    exclusion and 128

experimentation 72–3
fireplaces 61
gentrification and 121, 161
kitchens 63–4, 67, 72, 74
large-scale projects 126
memories 46
mistreatment and 97, 98
orchards 18
purist approaches 73, 126
scope 98, 108, 124, 125
stewardship 70
trust 59
windows 44, 59
MOHPG (Monrovia Historic Preservation
    Group) 123–4, 125, 126–7
Monrovia Historic Preservation Group
    (MOHPG) 123–4, 125, 126–7
Morris, W. 81, 82, 83
Mount Rubidoux 90–1
Mount Vernon Ladies' Association of the
    Union 83–4
Murfreesboro (TN) 27–8

National Historic Preservation Act
    (NHPA) 84
National Register of Historic Places 84,
    100
neighbors 3, 35, 91–2, 104, 116, 134
    anecdotes 39
    bonding and 2–3, 91, 113–14, 135
    exclusion and 114–15
NHPA (National Historic Preservation
    Act) 84
nostalgia 26–7

oak 31f.
Oak Streets 91, 112, 113–14
    challenges 93
    crime and 112–13
    disengagement and 111, 115
    disparity 91–2, 113, 114
    districting and 90
    exclusion and 92, 113, 114–15, 116
    parking issues 111–12
    scope 91, 92–3, 116
Old Riverside Foundation (ORF) 109, 110
    disparity 110
    events and tours 110–11
    exclusion and 90

limitations 88–9
Ontario 18–19, 48, 93–4, 120, 162
    costs and 95
    de-listing and 94, 117, 118
    disengagement and 119
    disparity 94–5, 108, 109, 118–19
    districting and 94, 95f., 117
    exclusion and 119–20, 161
    experimentation 71–2
    heirlooms 47
    limitations and 116–17, 118
    memories and 46–8, 75, 77
    mistreatment and 119
    orchards 18
    purist approaches 74–5, 77
    scope 93, 108, 117
    stucco and 117, 118
    trust and 59
Ontario Heritage 116–17, 118–19
orchards 1–2, 18
ORF, *see* Old Riverside Foundation
originality and authenticity 3, 11–12,
    13–14, 33
    disparity 48
    essences and 12
    limitations 12

paint 38–9, 61, 68
    costs 55
    experimentation 53
parking issues 111–12
Pasadena 18–19, 27, 35, 36, 42, 70, 96,
    135, 162
    air conditioning 60
    bathroom 43, 57–8
    carbon monoxide leak 44
    challenges 99–100, 132, 133
    coop 37
    demolition and 99, 133
    disparity 27, 100, 134, 137
    districting 98–9, 100, 130
    events and tours 130–1, 131f.
    experimentation 52
    fireplaces 61, 62, 68
    floors 52–3
    gentrification and 121, 128–9, 132,
    135–6, 161
    ghost marks 44
    kitchens 62–3, 64–7

large-scale projects and 129–30, 132–3
light switches 70
memories 27
mistreatment and 40, 64, 131–2
newspaper in wall 34
orchards 18
original style and 34–5
paint 38–9, 53
picture rails 68
purist approaches 56, 67–8
scope 98, 108, 129, 130, 134–5,
    136–7
stewardship 70–1
stucco and 39–40
trust 59
wainscoting 35
wallpaper 68
weekend work 133–4
windows 43
Pasadena Heritage 129
patriotism 84
picture rails 68
porches 65–6
Portland (OR) 166
privacy 5
profits 108, 132
Pueblo Revival property 70
    fireplaces 61, 68
    kitchens 64–6
    light switches 70
    purist approaches 67–8
    stewardship 70–1
    trust 59
purist approaches 51, 56–7, 57f., 62, 67–8,
    73–4, 77
    challenges 51, 57–8, 62, 68
    disparity 60–1, 74–5, 77–8, 126, 160
    stewardship 57, 58

Queen Anne property 48, 75f., 76f.
    deck and 125
    memories 47–8
    purist approaches 74–5

RCC (Riverside Community College)
    111–12
redwood 40
refrigerators 7–8, 57f., 63
rental property 36–7, 108, 113, 151

Riverside 18–19, 91, 110, 112, 113–14,
    120, 162
    challenges 89, 93
    crime and 112–13
    demolition and 88, 109–10
    disengagement and 111, 115
    disparity 91–2, 108, 109, 110, 113, 114
    districting 90
    events and tours 110–11
    exclusion and 90–1, 92, 113, 114–15,
        116, 119–20, 161
    kitchens 73
    limitations 88–9
    orchards 18
    parking issues 111–12
    purist approaches and 77–8
    relocation of property 110
    salvage and 110
    scope 89–90, 91, 92–3, 108, 109, 116
Riverside Community College (RCC)
    111–12
Riverside Renovators 111
roofs 154–5
Rosewood Place 91
Ruskin, J. 81, 82–3

salvage 110
"Secretary's Standards" 87
shingles 39–40, 41f., 42f., 151
shiplap 57f.
Smith, N. 15
Spanish Revival property 4f., 47, 149–50,
    154
Speicher, I. 34
Standards for Rehabilitation 87
stewardship 50–1, 57, 58, 70
    challenges 70–1
Stickley designs 75, 77
stoves 63, 66
stucco 39–40, 41f., 42f.
    mistreatment and 117, 118, 147
sweat equity 23

taxes 88
territoriality 5–6
Theodossopoulos, D. 12
tiles 43, 61
Tilley, C. 8
toe-kicks 65f.

tract property 32
trust 59
Tudor Revival property 126–7
   knowledge of styles and 149
   large-scale project 153
   roofs 154–5
Tyler, N. 85

Veblen, T. 83
Victorian property 48, 110
   demolition and 97
   doors 73–4
   kitchens 74
   memories 46
   mistreatment and 97, 98
   purist approaches 73
   see also further styles
vinyl 151
Viollet-le-Duc, E. E. 82
voicefullness 13, 82

wainscoting 35, 72

wallpaper 68, 69f.
Walnut (CA) 32
wealth 140, 143, 156, *see also* elites
Wildrose 96–7
windows 43–4, 167
   carpentry 59
   ghost marks 44
   mistreatment 43
   purist approaches 56
   retrospective permits and 151
wood 38
   cabinets 28–9, 30f., 58–9, 72
   carpentry 54–5
   doors 31f., 32, 73–4, 167
   floors 52–3
   purist approaches 56, 57–8
   refrigeration and 57f.
   shingles 40, 151
   stucco and 117, 118, 147
   wainscoting 35, 72
   windows 59, 151, 167
Wright, F. L. 98